AF413455

A day at
CERN
Guided Tour Through
the Heart of Particle Physics

A day at
CERN

*Guided Tour Through
the Heart of Particle Physics*

Gautier Depambour

Université de Paris, France

Illustrations: **Lison Bernet**

World Scientific

NEW JERSEY · LONDON · SINGAPORE · BEIJING · SHANGHAI · HONG KONG · TAIPEI · CHENNAI · TOKYO

Published by

World Scientific Publishing Co. Pte. Ltd.

5 Toh Tuck Link, Singapore 596224

USA office: 27 Warren Street, Suite 401-402, Hackensack, NJ 07601

UK office: 57 Shelton Street, Covent Garden, London WC2H 9HE

Library of Congress Control Number: 2020941449

British Library Cataloguing-in-Publication Data
A catalogue record for this book is available from the British Library.

A DAY AT CERN
Guided Tour Through the Heart of Particle Physics

ISBN 978-981-122-110-1 (hardcover)
ISBN 978-981-122-064-7 (paperback)
ISBN 978-981-122-065-4 (ebook for institutions)
ISBN 978-981-122-066-1 (ebook for individuals)

For any available supplementary material, please visit
https://www.worldscientific.com/worldscibooks/10.1142/11839#t=suppl

Desk Editor: Ng Kah Fee

Typeset by Diacritech Technologies Pvt. Ltd.
Chennai - 600106, India

Table of Contents

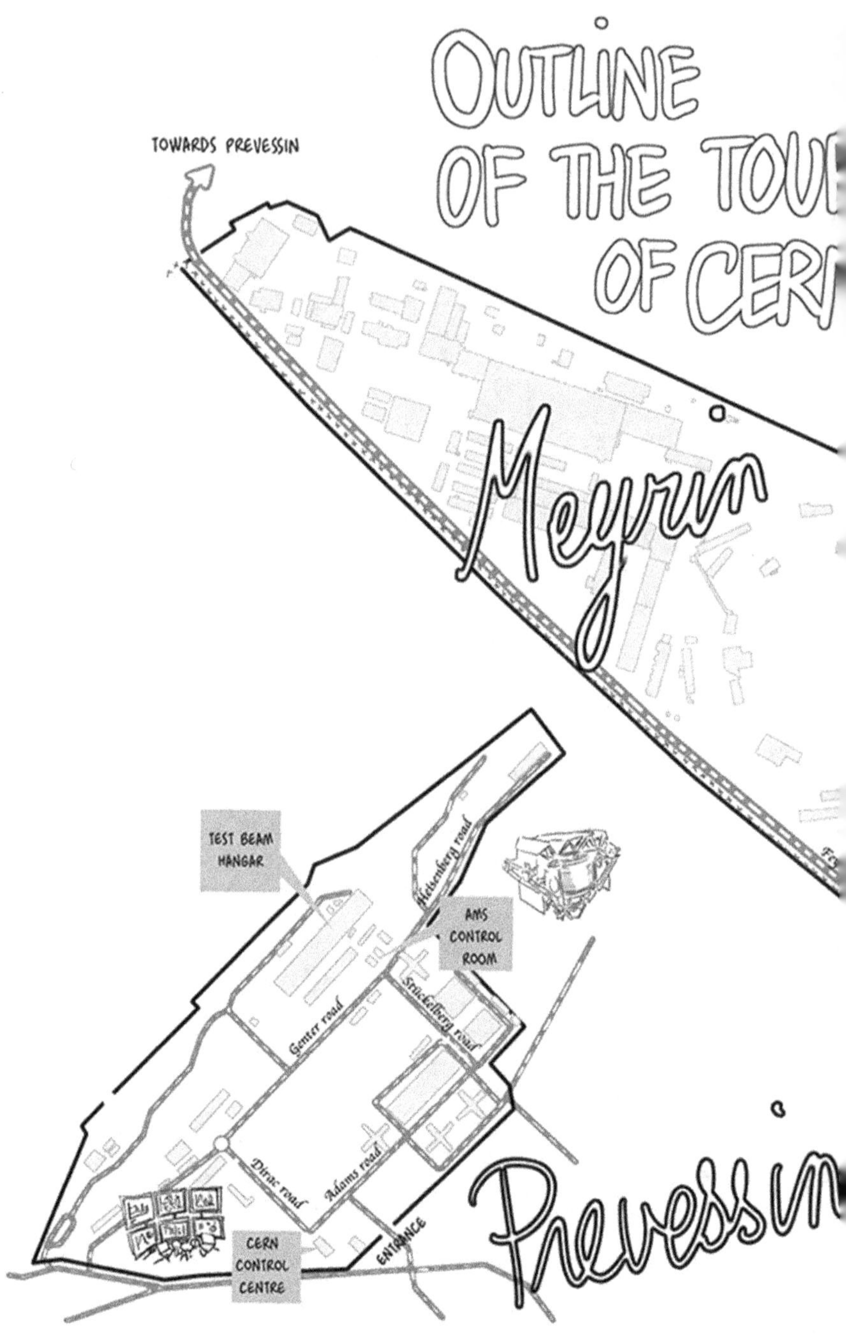

TOWARDS PREVESSIN
OUTLINE
OF THE TOUR
OF CERN
Meyrin
Prevessin
TEST BEAM HANGAR
AMS CONTROL ROOM
Heisenberg road
Schutenberg road
Gentner road
Dirac road
Adams road
ENTRANCE
CERN CONTROL CENTRE

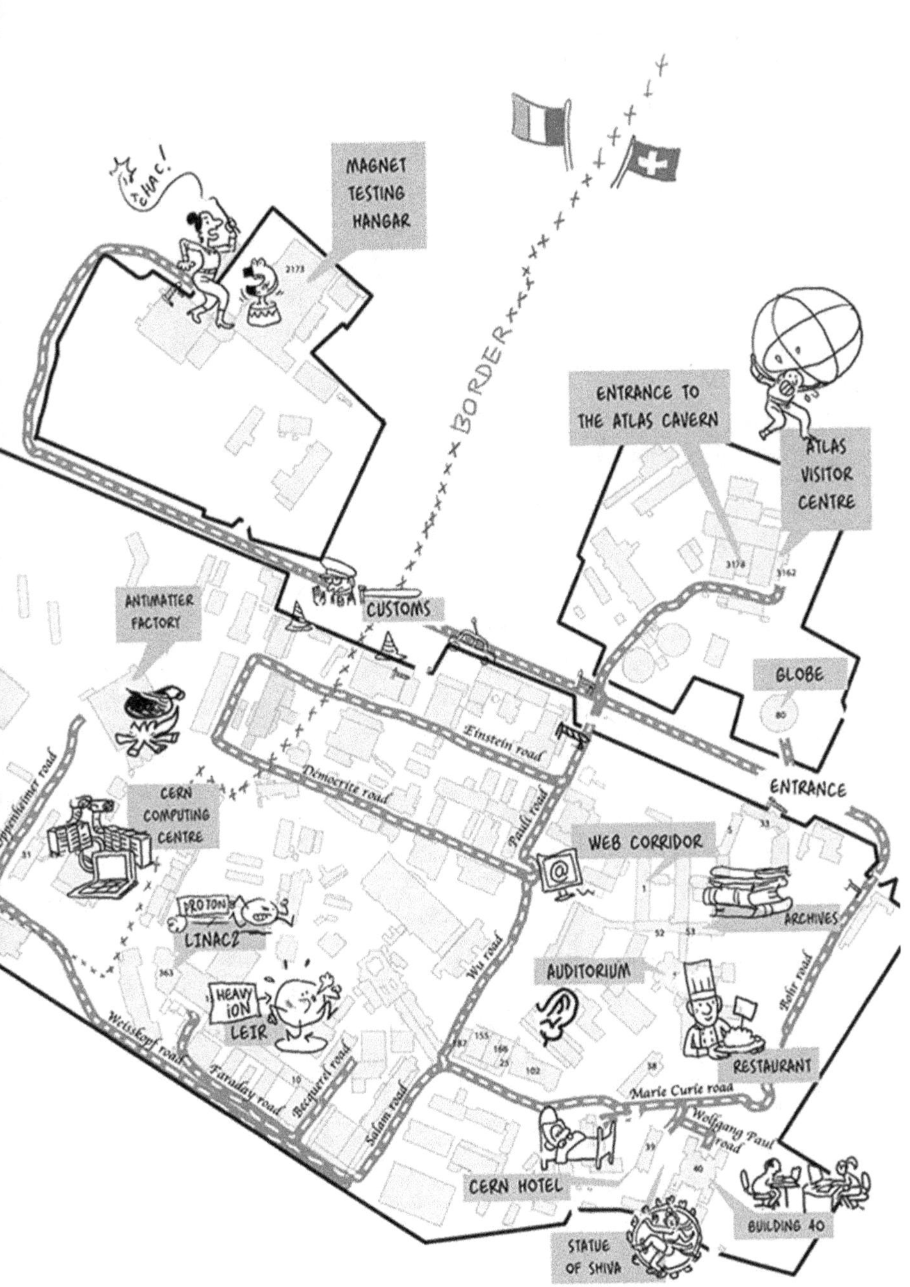
MAGNET TESTING HANGAR
ELHAC!
ENTRANCE TO THE ATLAS CAVERN
ATLAS VISITOR CENTRE
BORDER
ANTIMATTER FACTORY
CUSTOMS
GLOBE
Einstein road
Démocrite road
ENTRANCE
Oppenheimer road
CERN COMPUTING CENTRE
Pauli road
WEB CORRIDOR
ARCHIVES
PROTON
LINAC2
Wu road
AUDITORIUM
Bohr road
HEAVY ION
LEIR
RESTAURANT
Weisskopf road
Marie Curie road
Faraday road
Becquerel road
Salam road
Wolfgang Paul road
CERN HOTEL
STATUE OF SHIVA
BUILDING 40

Foreword

by Frédérick Bordry
Director for Accelerators and Technology at CERN

Writing a foreword is traditionally an opportunity to present ideas that are dear to you. When it comes to CERN, the word "dear" is weak. I joined it more than 30 years ago, and over the years since then, I have taken an intense, and perhaps even growing, pleasure in guiding people through this unique research centre. At first, the requests came only from private visitors, often colleagues from other laboratories, or relatives, curious or attracted by my already evident enthusiasm. I changed jobs about every 7 years, and so I got to know CERN better and better, as I experienced the events associated with it, and as I engaged in daily exchanges with the various participants in the project — technicians, physicists, engineers and others.

I had the privilege and pleasure of showing CERN to many representatives of Member and Associate Member States, as well as to various personalities. Here I will mention only Al Gore and Kofi Annan. I think of Al Gore because of his commitment to climate change and his interest in understanding fundamental research and its applications in the broad field of energy.

Kofi Annan was a great humanist and a tireless artisan for peace. He came to CERN and returned to it to praise science for peace and development. He was always interested in the latest technology transfers and in particular those that could help Africa's development.

Different cultures, different languages, and different professions make CERN a place of tolerance and peace; this point is underlined several times in this book, but how can I not highlight it, too?

This *Day at CERN* is written like an adventure novel: it is one. For more than 60 years now, many books, brochures and other articles have been written to explain what is being done at CERN, but Gautier Depambour's has many special features that make it attractive. It is written in a light, airy style. The sentences are well formed, and an underlying humour is always present. A few anecdotes, always of high quality, will make you smile. The text is that of an adventure, as I said, but a shared adventure. The greatest originality of this story is that it is addressed to you — yes, you — personally. You are present throughout the visit: you are invited in; you are given a badge; you are introduced to the people who will give you their vision of CERN; you are given an explanation of some points that you may find unclear.

As an engineer, I have noted the author's "physicist" vision. I think it is worth mentioning that among CERN's 2500 international civil servants (permanent or not), the vast majority are technicians, engineers and applied physicists; less than a hundred are theoretical physicists! All their efforts are dedicated to the design, installation and proper functioning of large instruments and infrastructure for the world's 13,000 physicist users.

Lison Bernet's illustrations bring another touch of freshness. All those who are asked to talk about CERN and what they do there are well known to me, and it is a pleasure to see them shown in such a personal way. The particularities of each offer many opportunities to add a touch of humour, as with, for example, the photo of John Ellis' famous office.

The whole is surprisingly precise and rigorous, showing a concern to comment on all kinds of details: technical installations of course, but also street names and cafeterias (in particular the one I call Babel-cafeteria for the number of languages spoken there), with the names of the dishes offered there... The tour includes the Innovation Globe, the reception hall and various other facilities. After reading this, will you really need to come to CERN? I will bet that you will have an even greater desire to go there, during the Open Days that we organize every 4 years, or during one of the visits organized every Saturday upon prior registration. And if, on the contrary, you already know CERN, you, like me, will enjoy seeing it from the angle chosen by Gautier Depambour, whom I congratulate on this impressive work.

Frédérick Bordry

A Welcome at CERN's entrance

I'm here with the ATLAS cap on — hello! And welcome to CERN's entrance. You arrived as planned from Geneva, by the tramway. I am delighted by the curiosity that has led you to come here, and I am pleased to welcome you to the Franco-Swiss border to show you this exceptional place dedicated to particle physics. I will be your guide throughout the day.

I will give you an overview of the research being carried out at CERN, the "Conseil Européen pour la Recherche Nucléaire," renamed in 1954 the "European Organization for Nuclear Research," although the old acronym has been retained. I will of course mention the famous Higgs boson, and I will show you how the world's largest particle accelerator works, the LHC (Large Hadron Collider), located about 100 m below our feet, as well as one of the four detectors where particle collisions occur: ATLAS.

But even more importantly, I intend to immerse you in CERN's daily life in order to make you feel the atmosphere of this place; you will see it in a way completely out of the ordinary. During my five-month internship in the ATLAS detector communication group, I had the opportunity to meet some very nice people, and I now want you to benefit from this. That's why I have organized several meetings with people who work in various fields, from accelerator physics to artificial intelligence used for data analysis. Today, I would like to share with you my enthusiasm for CERN and research in high-energy physics as it is carried out there!

Well, let's take a look around us. First of all, you will have admired this imposing wooden structure: it is the Globe of Science and Innovation, which has become CERN's emblem over the years. In addition to a large conference room, it houses a magnificent exhibition, both informative and poetic, entitled "Universe of Particles," which allows one to discover what research at CERN is all about — and makes it a must for visitors.

The Globe was built in Neuchâtel for the 2002 Swiss National Exhibition and symbolized sustainable development, because wood behaves like a carbon sink and thus reduces the greenhouse effect. At the end of the Exhibition, the future of the Globe was the subject of reflection. There was no question of burning it — that would have been inappropriate for a building supposed to represent sustainable development. Finally, the Swiss Confederation decided to offer it to CERN on the occasion of the Organization's 50th anniversary in 2004. That's why it ended up here! However, the Globe in front of you is not completely identical to the original: it was necessary to change the outer arches, which had not initially been made of wood that could withstand decades. Now reinforced, this building will continue to accompany CERN's history throughout this century, while remaining a symbol for Geneva and its region. Now, let's turn around...

The Globe of Science and Innovation

Flags at CERN's entrance

Do you see this line of flags? They correspond to all the member countries participating in this immense scientific collaboration. Could you guess in which order the flags are arranged? Answer: In alphabetical order, according to the names of countries in French. So the line starts with Germany and ends with... Switzerland, as you can see.

CERN was initially a European project, whose Convention was signed by 12 founding countries in 1954. By 2016, this number had risen to 22 Member States — including one that is not European, Israel — and six Associate States: India, for example, joined CERN as an Associate State in December 2016. An Associate State does not have the same status as a Member State: it contributes financially in the same way, and, above all, does not have the right to vote in the CERN Council, the highest ruling body of the Organization, where important decisions are taken. Finally, there are universities that have cooperation contracts with CERN. In all, about a hundred nationalities are represented here: that is quite exceptional.

First, let's head to building 33: this is the reception area, where visitors are welcomed.

CERN's entrance for visitors

As we walk, let me give you an overview of the day's programme — without revealing everything, of course.

In order for you to understand our journey, one thing must already be clear. Do not confuse detector and accelerator! 100 m underground is a circular accelerator with a circumference of 27 km, the LHC, which accelerates particles. And at four different locations on this accelerator are arranged the four main detectors where particles collide. Their names? ATLAS, CMS, ALICE and LHCb. Keep in mind for now this distinction between the LHC accelerator and the four large detectors at the collision points.

I now come to the programme for the visit. First, by way of introduction, we will go to the Globe to visit the exhibition I was talking about a moment ago. Then, our journey through CERN's various facilities will follow the same path as protons do, from their injection into accelerators to their collision inside detectors. Then, our visit will follow the logic of data analysis, from the acquisition of raw data by a detector to the discovery of a new particle.

But before getting to the heart of the matter and getting to the Globe, I imagine that you are spontaneously asking yourself some questions about CERN, such as: what research is being done here? What is all this for, and why should I feel conCERNed? How many people work at CERN, and in what professional capicities? To answer all these questions, I have arranged a meeting with Bernard Pellequer, who is responsible for the Globe exhibition in particular. We'll meet him in his office soon.

We have arrived in the reception hall: nice, isn't it? All the people who work at CERN can become guides, and that makes the visits very lively and authentic. Follow me to the desk on the right to pick up your visitor's card; you can already take out your ID card.

"Hello madam, I ordered a pass on Indico... Yes, that's it, thank you." Here's your card. For my part, I have to get my badge out if we want to get through that glass door. I always thought the picture of me looked strange; I have asked several people who I reminded them of. I have been given answers that I am still trying to analyze: Michael Scofield in *Prison Break*, Robert De Niro in *Taxi Driver*, Mark Ruffalo in *Avengers* before becoming Hulk (fortunately not after), and even a lifeguard, which is in my case a remarkable misunderstanding.

I won't hide from you the fact that I'm rather proud to own this badge. For you, it is a small piece of cardboard valid for one day with your name on it — and mine, because I am responsible for you. Be aware that "normal" visitors, if I may put it that way, do not even have this pass. With this, you are privileged, and you can legally access a large part of CERN without having to be in a group accompanied by a guide. In other words, you are freer, even if you depend on me!

Let's go through that glass door. That's it, we're going into CERN's inner sanctum. Here we go. Like a guide opening the hidden door of a library leading to the secret cabinet of an old castle, I am taking you into the nooks and crannies of the world's greatest scientific collaboration...

Ah, stop! Not too fast. I can see that you are eager to go further, but we will stay for a while in building 33. Bernard's office is there, just on the left. I think he's coming to open the door, he must have seen our silhouettes through the tinted window of his door. If I need to talk to you discreetly during the interview, I will whisper. There he is.

My access badge

Bernard Pellequer

Interview with Bernard Pellequer

Presentation of CERN

"Hello Bernard! Let me introduce you to the person who is accompanying me throughout the day, as I explained to you.

- Hello, both of you! Please, come in and have a seat.

I'd like to start by asking you a very simple question... What is the purpose of CERN?

- There is an obvious answer: it is a laboratory that brings together a group of people who will contribute — with all the modesty behind this expression — to the increase of humanity's knowledge about the universe in which we live and about the structure of matter. That is the primary mission. But, of course, we cannot limit ourselves to that! Behind this scientific quest, there is a duty: to have technologies adapted to this research. And sometimes these technologies are transferred to industry or the general public.

This leads me to a second question that is just as crucial as the first: what is the purpose of CERN for us?

- To stimulate our curiosity! Research on the universe and matter contributes to global knowledge. In addition, CERN also serves us all indirectly in our daily lives. For example, in the medical field, Magnetic Resonance Imaging (MRI), Positron Emission Tomography (PET), which is used in oncology and neurology, and all the techniques employing radioactive tracers are based on technologies developed for research. The analytical equipment is simply miniaturized and adapted to human morphology. And for treating patients, we use accelerators that also take advantage of the advances made at CERN.

But then, does that mean that there are accelerators all over the world?

- Absolutely! There are those dedicated to research, such as those at CERN, of course; but throughout the world, more than 30,000 accelerators are used in very different fields, in medicine and even... at the Louvre! The museum has an accelerator that makes it possible, for example, to determine the chemical composition of artworks, to date them, and sometimes to confirm their authenticity."

This is a rather unexpected application!

"Let's talk a little more about CERN, Bernard. How many people work here?

- Currently, there are 2500 people working directly for CERN, about a thousand students and doctoral students also paid by the Organization, as well as about 13,000 users from about 100 countries, who are paid by their home laboratories and who come periodically for missions. Some stay for a few days, others for a few weeks or more. The right question to ask, therefore, is: How many people are there at CERN on a given day? And the answer is: between 7000 and 10,000, depending on the operating periods of the accelerators.

And are all of these scientists immersed in their calculations?

- Oh no, far from it! It should be pointed out that there are not only scientists at CERN, a large number of professions can be pursued here. Many different training courses allow you to work at CERN — as a theorist, engineer, technician, assistant in an administrative field, member of the service staff, etc. There is a continuous spectrum of very varied jobs! And of course, the selected candidates are the best in their fields."

The selection must be rigorous. Well, let's get to the Higgs boson.

"Bernard, the LHC was built mainly to discover the Higgs boson... Now that this objective has been achieved, what are we going to do?

- Well, first of all, the Higgs boson story is far from the end! The coming years of operation of the LHC will serve to refine our understanding of the Higgs boson itself, but also to study the couplings that exist between it and the particles that acquire mass through this interaction. And of course, in parallel, we will continue to test the Standard Model..."

I don't know if all this means anything to you, but in any case, don't panic. We will obviously have the opportunity to discuss it again during our visit.

"... At the same time, we will look carefully to see if there might be shivers of other particles that we do not know and that cannot be predicted by any theory. Between its inauguration and 2016, the LHC provided only 5% of the total number of collisions planned, with a research programme that extends to 2037. The LHC therefore still has a long way to go. We hope to be able to provide answers to some of the major questions of current physics: what is matter made of? Why did matter prevail over antimatter in the early universe? What is dark matter? We don't have an answer yet, but we have avenues to explore, especially in particle physics. Thus, it is by observing the smallest that we can understand the biggest!

Speaking of smaller and bigger... Can the LHC produce black holes, as in space, but on a very small scale?

- This is very unlikely. However, according to some theories, the formation of tiny "quantum" black holes is possible. Observing such a phenomenon would allow us to better understand the universe... and would not present any danger! We are constantly receiving particles from space that have an energy much higher than the LHC can produce, and that collide with particles in the atmosphere. Does this mean that these collisions lead to black holes? No. So how could this be the case in the LHC, with much lower energy?"

So there is no danger — the Earth will not disappear into a black hole created at CERN!

"Well, thank you very much Bernard, I think that's all that...

- No, wait! Wait! Before you leave, I would like to add something that is close to my heart. I would like to tell you that CERN, for me, is not only a research laboratory: it also represents more than 60 years of science in the service of peace..."

CERN was born just after the Second World War: later on, I will tell you a little bit about its history.

"... Only CERN has carried and continues to carry this message of peace with the same strength. For example, the SESAME project (supported by CERN and UNESCO) for a new synchrotron accelerator in Jordan brings together Israelis and Palestinians, but also Egyptians, Iranians, Pakistanis, Turks, Cypriots, Bahrainis and of course Jordanians. Symbolically, it is extraordinary: I am firmly convinced that we can unite countries through science. CERN is a model of international collaboration, which promotes respect for cultures and origins, and which, through its ability to bring together very different people, provides an answer to the question: how can we live well together? In my opinion, this aspect is very important.

So we will leave each other with this cry from the heart! Thank you, Bernard, for sharing your knowledge of CERN with us.

- You're welcome; if you have any further questions, feel free to come back to my office."

After this warm and informative nice introduction offered by Bernard, I have only one desire: to take you to the Globe! Let's go out the way we entered — it's the fastest. You will tell me that we have not discussed economic issues with Bernard: don't worry, we will have another interview very soon that will be specifically devoted to that question.

As we walk along the impressive line of flags, you may wonder what languages are spoken at CERN. There are two official languages here, English and French; yet, against all odds, neither is the majority language. Don't be surprised: you can rest assured that no one here speaks Shakespeare's language, or even *like* Shakespeare; no, here the unofficial but nevertheless the most common language is globish. For French people like me, it is, if you will, a generalization of "Frenglish" to the whole world, with a rather limited vocabulary. So, what you need to understand is that here we do not speak all the languages, but that we speak one — globish — with all possible accents.

There is one word in particular that you should know in globish, *plot*. Here, when you see what looks like a histogram, a graph, a diagram or something similar, you don't look for the precisely appropriate word, you just say: plot. This can save you from tricky situations, such as when you start an emphatic sentence with: "Oh! Very nice, your... your uh..." And then, since you don't know the exact word, you let the worst escape: "...your drawing." So if you have any doubts about what is represented, do as everyone else does; don't take any risks and say "plot."

With plot, you already reach the semi-pro grade, but if you want to appear immediately as a true professional, I have a trick to suggest: season all your sentences with *basically*. "Basically" is a word that, in English as in globish, is meaningless but very effective in making you shine. If you are asked a question, especially if it concerns particle physics, assume a very inspired look, and start by saying in your most beautiful globish "Basically, I think that...." You will give considerable weight in advance to your words, I assure you.

We're arriving at the entrance of the Globe — after you.

The exhibition "Universe of Particles"

As you will see, the effect is quite striking when you discover the exhibition room. Impressive, isn't it? I am always amazed by the beauty of this room. I particularly appreciate its cosmic aspect: in the middle of these big luminous spheres, one feels projected into interstellar space.

The exhibition "Universe of Particles"

This room thus sends to everyone an essential subliminal message, which Bernard has already mentioned earlier and which I find magnificent: it is by observing the smallest things that we can understand the biggest. What a beautiful idea! Perhaps we will find answers to astrophysical mysteries, such as dark matter or dark energy, by studying the smallest components of matter. At CERN, therefore, we do not limit ourselves to understanding the infinitely small: we try to understand the universe on every scale.

The spheres you see in this room — at least those close to the ground — each contains a curiosity relating to CERN's history, to accelerators or particle detectors, or to a point of theoretical physics. We're going to discover some of them, but just before that, I have to take out my secret weapon: My beautiful drawing of CERN's accelerator chain! With this visual aid, I will be able to describe the path of the protons: after being accelerated, some of them will frontally meet other protons coming from the opposite direction. The energy released during such a collision will allow, by virtue of the equivalence between mass and energy formalized by Einstein, the production of new particles that can be studied using particle detectors.

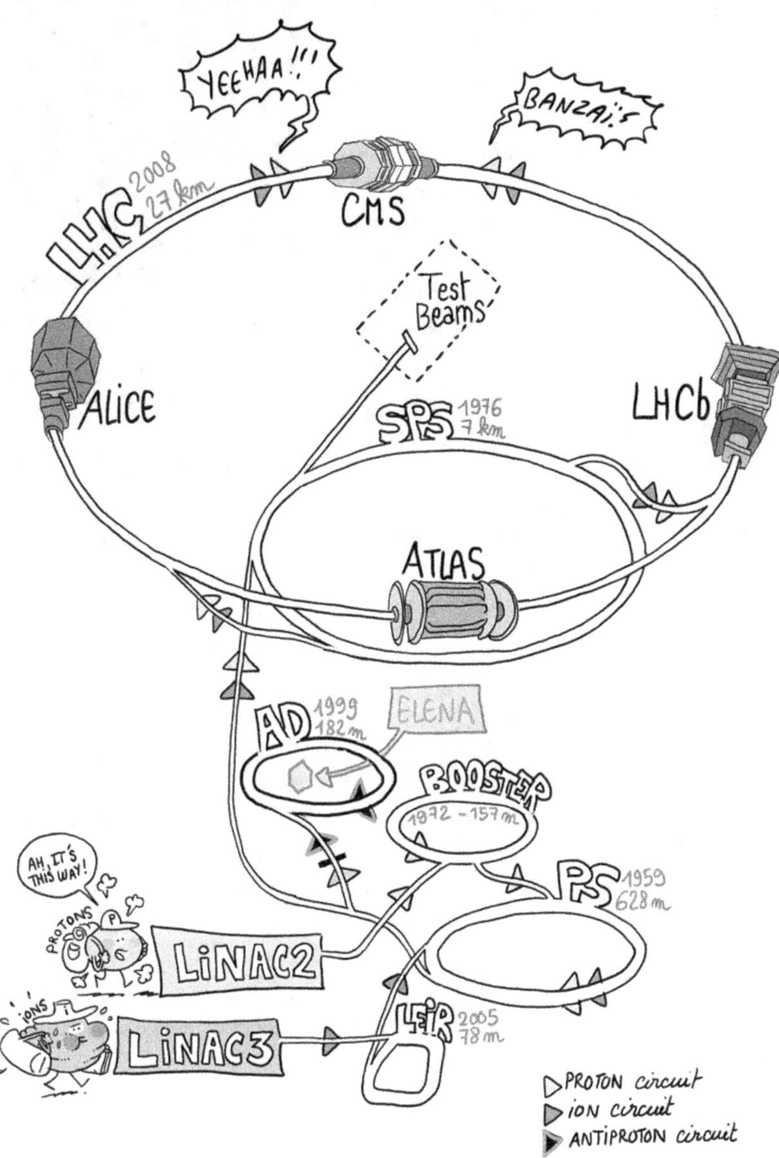
THE CERN ACCELERATOR COMPLEX
YEEHAA!!!
BANZAI's
LHC 2008 27 km
CMS
Test Beams
ALICE
LHCb
SPS 1976 7 km
ATLAS
AD 1999 182 m
ELENA
BOOSTER 1972 - 157 m
PS 1959 628 m
AH, IT'S THIS WAY!
PROTONS P
IONS
LiNAC2
LiNAC3
LEIR 2005 78 m
PROTON circuit
ION circuit
ANTIPROTON circuit

But let's not go too fast, and let's start at the beginning: where do these protons come from? Don't expect a staggering answer, because the protons simply come from a hydrogen bottle. This is an opportunity to show you the first sphere — come closer.

Hydrogen is the simplest atom that can be found: its nucleus consists of only one proton of positive electrical charge +1, around which gravitates an electron of negative electrical charge −1. More precisely, if we do not want to stick to this classical image, which dates back more than a century, it is better to imagine the electron as an electronic cloud which represents its probability

of presence around the nucleus. However, there is no need to discuss the electrons, since they are separated from their respective protons as soon as they leave the hydrogen bottle, thanks to an electric field.

The protons thus obtained are first injected into a straight-line accelerator: LINAC 2, which stands for LINear ACcelerator 2. Its cousin, LINAC 3, which you can also spot on the diagram, accelerates not protons but Pb^{29+} lead ions, i.e., lead atoms that have been stripped off some of their electrons. However, most of the time, accelerated particles are protons — let's focus on them for now. They will go through a succession of increasingly powerful accelerators, which will make them go faster and faster, giving them more and more energy.

Why do we need several accelerators in a chain, and why don't we send the protons directly into the largest accelerator? For a simple reason: because you don't immediately switch from a local road to the highway. It is as if you were asked to arrive on the highway at 40 km/h, then accelerate to 130 km/h: this is not possible; cars that are already there are going too fast. The protons therefore pass through successive insertion ramps. In the life of an accelerator, there is no retirement: when you are supplanted by a bigger one, you become the insertion ramp of the newcomer.

First, in LINAC 2 — note this! — protons are grouped in *packets*. Then these packets of protons are sent to the Booster, which accelerates them and sends them to the PS — the Proton Synchrotron with its 628 m of circumference. The PS (which is located on the surface) accelerates them and then sends them 40 m underground to the SPS, the Super Proton Synchrotron, with its 7 km of circumference. The SPS accelerates them and then sends them to the LHC, the Large Hadron Collider, with its 27 km of circumference. But here there is a subtlety: the SPS sends half of the proton packets clockwise into the LHC, and the other half counterclockwise for the purpose of producing collisions!

As I said at the outset, the LHC is now the largest and most powerful accelerator in the world. There are even more ambitious projects under consideration, which we will discuss at the end of the day in offering a perspective on the future, but which remain at an embryonic stage for the time being. At the moment, collisions produced in the LHC reach an energy of 13 tera-electronvolts. If this number doesn't tell you much, just consider that the energy contained in a particle at full speed in the LHC is equivalent to that of a mosquito in flight — except that at CERN, this energy is concentrated in a tiny proton, not distributed among the billions of billions of protons that make up the mosquito!

Collisions occur in the four detectors I mentioned earlier: ATLAS (A Toroidal LHC ApparatuS), CMS (Compact Muon Solenoid), ALICE (A Large Ion Collider Experiment) and LHCb (Large Hadron Collider beauty). CMS research, like that of ATLAS, focuses largely on the study of the Higgs boson. The ALICE experiment is intended for the study of quark–gluon plasmas, a kind of primordial magma that, by cooling, gives rise to other particles of matter. To do this, collisions within ALICE involve heavy ions, i.e., large chemical elements; there are periods when the LHC doesn't provide collisions between protons, but between much heavier elements, such as lead. The LHCb experiment, for its part, focuses on the quark b, known as "beauty" or "bottom" — a fleeting particle (because its lifetime before decay is very short) that helps physicists to understand the differences between matter and antimatter. All these detectors have different technologies adapted to their research programmes. At the risk of repeating myself later, I would like to stress that ATLAS and CMS, although having very similar programmes, do not use the same technologies for particle detection, or, therefore, the same methods. This is very important, because observing the same phenomenon with two different detectors ensures the reliability of the results and limits errors.

The detectors record the data from the collisions, which are then processed, in comparison with the simulated data, until an experimental result is announced, such as the discovery of a new particle. But one thing at a time: so far, the important thing has been to give you an overview of the path of a proton from the hydrogen bottle to the various detectors of the LHC.

One remark must be made here: this path is in fact only used by less than 0.08% of the protons! It should not be forgotten that many other experiments take place between the Booster and the LHC. For example, some accelerated protons in the PS, instead of being injected into the SPS, leave for the antimatter factory, which we will visit later; others leave for a lead target, producing high-energy neutrons studied by the n-ToF

The tunnel of the LHC

experiment, with applications in both astrophysics and medicine. The same is true for the SPS: some protons, if not injected into the LHC, are directed to the AWAKE experiment, which employs another plasma-based particle acceleration technique; others leave for the COMPASS experiment, which studies quarks and gluons within the atomic nucleus. So, even if we focus today on the ATLAS experiment, there are many others.

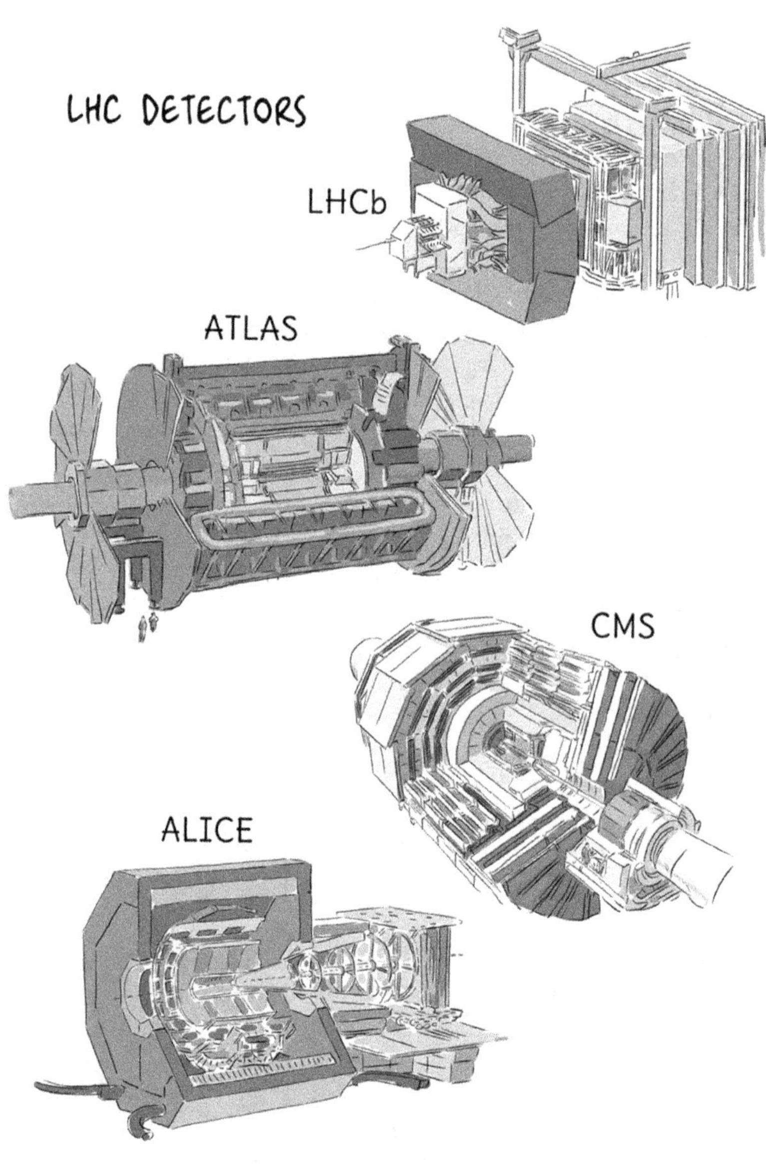

Follow me as I move to the other side of this room in the Globe. I would like to show you a little curiosity — which might even be raised to the status of an object of worship. Look into that sphere.

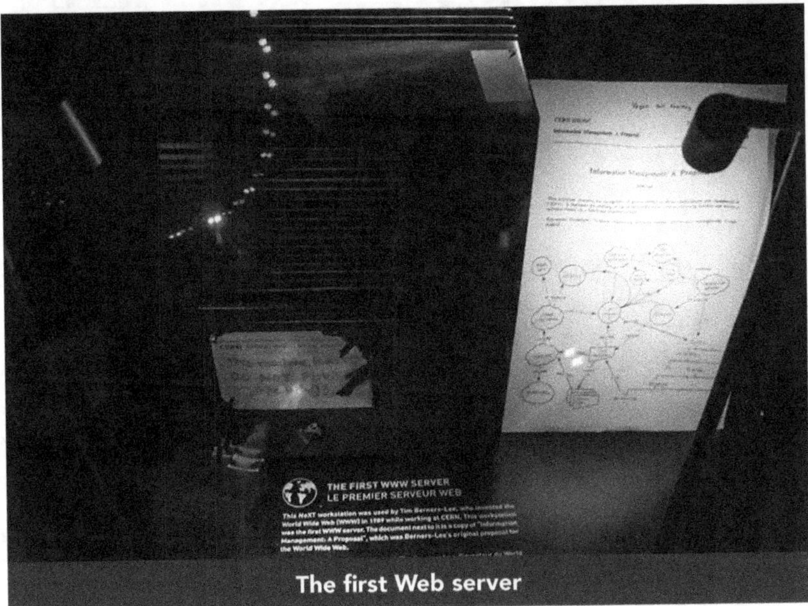

The first Web server

This is the first Web server, which was developed by Tim Berners-Lee in 1989. Take in what I'm about to tell you: the World Wide Web and the HTTP transfer protocol were invented at CERN! If you still had doubts about the usefulness of CERN, keep repeating to yourself: without CERN, no Facebook, no Google, no YouTube, no Wikipedia, no Amazon, no Twitter... Imagine your life today without the Web. Well, that would be a life without CERN.

Next to this server, you can see a document: it is the article that Tim Berners-Lee submitted to his superior in presenting his project. The reaction of the superior: "Vague... but exciting!" Initially, the Web project was intended to allow data sharing between physicists: we know what it is today.

But the Web is not the only contribution of CERN to IT! It is also at CERN that the idea of the touch screen was invented. So, once again, imagine a world without a touch screen: it's a world without CERN. Last but not least, it is at CERN that the prototype of what can be considered as the ancestor of the computer mouse was developed. That was in 1972, a certain Bent Stumpe had to design a system for the SPS control room — which I was telling you about a moment ago — to move a cursor on a control screen. He then ordered 12 bowling balls and developed the precursor of the ball mouse. The bowling ball was chosen for reasons of stability and fluidity of movement, but fortunately, all this could be miniaturized.

To anyone who would tell you that CERN has never been used for anything and that it is expensive, I now count on you to tell him or her that without CERN there would be no

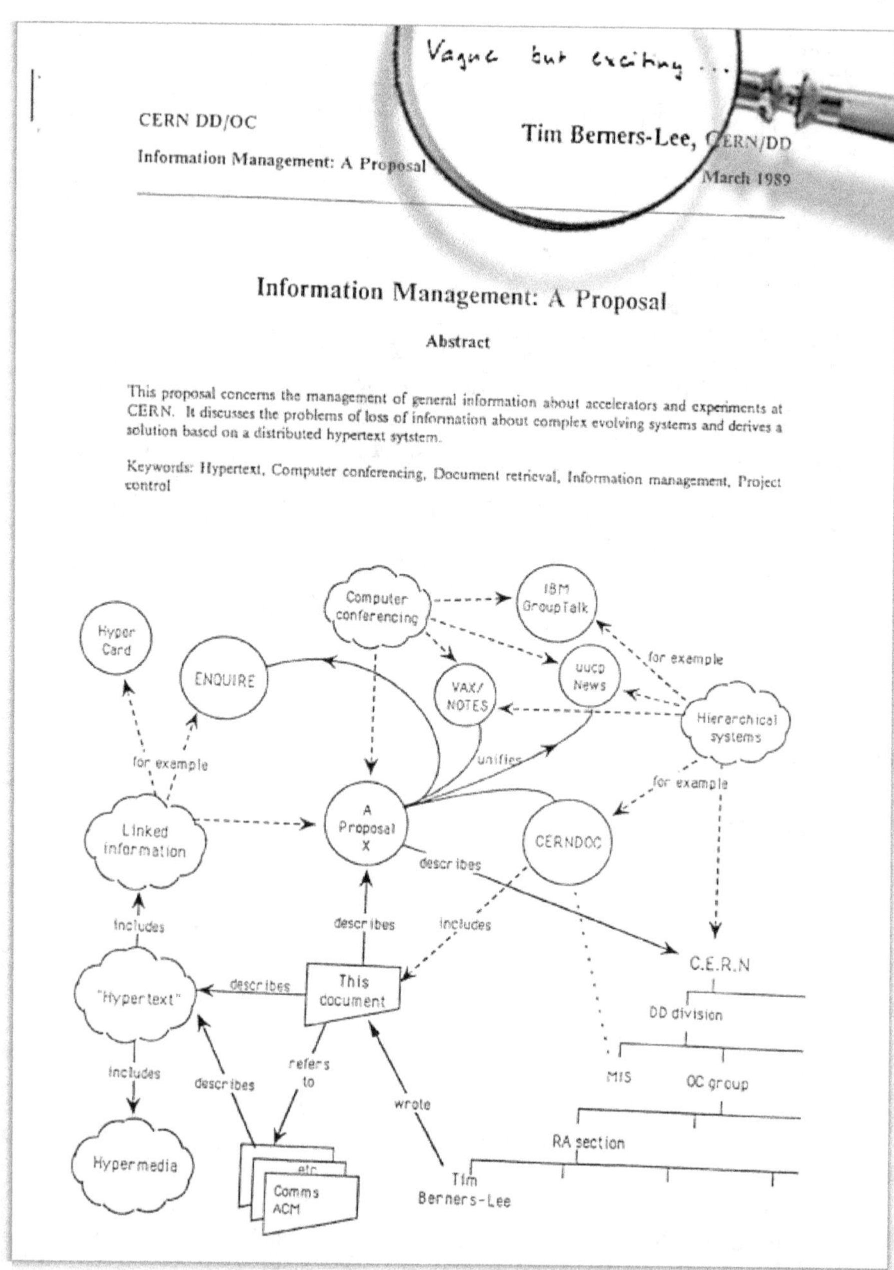

The article at the origin of the Web (Copyright: CERN)

Web, no touch screen, no computer mouse. And if this person persists in his scepticism, sincerely, it is because he or she is acting in bad faith.

Before leaving the Globe, I have two more little surprises. Come on...

The first particle accelerator

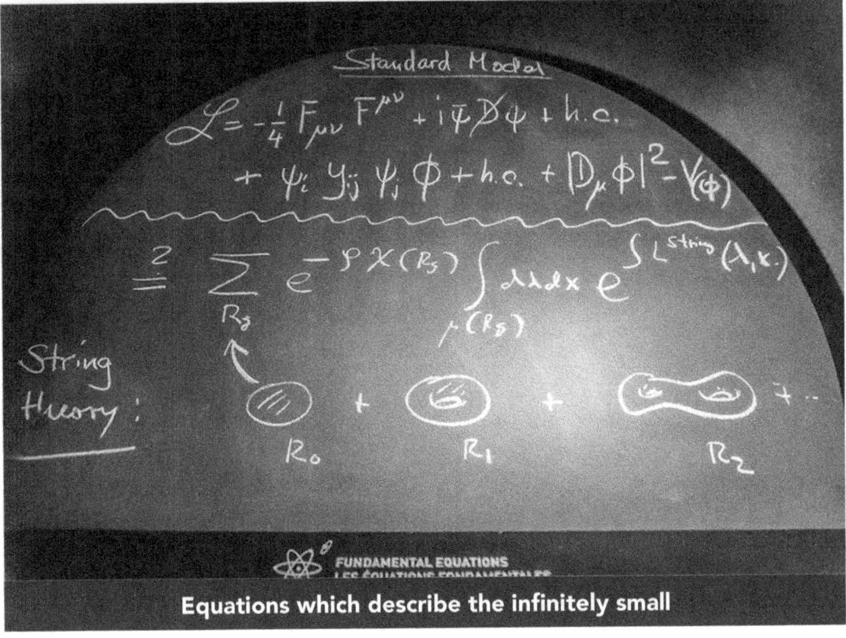

Equations which describe the infinitely small

In this sphere, you can contemplate the very first circular particle accelerator, a few centimetres in diameter. Don't you find the experience moving? And in this other sphere, here, you can admire the beauty of the equations that rule particle physics today. There's no rule against getting a little bit ecstatic!

Let's leave the Globe. Now you're ready for the discussion we're going to have with Arnaud Marsollier. Arnaud is part of the communication department, and I made an appointment with him to talk about money issues: how much did the LHC cost? And the detectors? What are the various economic and societal benefits? I think it is useful to understand from the beginning of the visit that the money invested in CERN is not lost money, far from it. But first, on the way to his office, let me tell you a little bit about the history of CERN, so that you can feel the weight of history as you walk through its corridors...

A brief history of CERN

If there is one place in the world devoted to science that beats records, it is CERN! Imagine: an international collaboration with 22 Member States in 2016 has built a ring of 27 km in circumference, at an average depth of 100 m, to cause tiny constituents of matter (such as protons) to collide at a vertiginous speed almost equals to that of light. When you work at CERN and repeat this description almost mechanically, you tend to get used to it and forget that... it's just amazing! But behind such exploits lies a whole history, whose story deserves to be told.

As you may know, the first half of the 20th century deeply changed our view of physics: first with Einstein, who formulated his theories of special relativity and general relativity in 1905 and 1915 respectively, and then with the rise of quantum mechanics, which developed considerably from the 1920s onwards. However, all these conceptual revolutions did not take place just anywhere: they took place in Europe. Heisenberg, Pauli, Dirac, Schrödinger, de Broglie, Born, Bohr, Einstein: almost all the great physicists who founded quantum physics were European.

But the Second World War, which led to a brain drain to America, put an end to this golden age, and fundamental research in Europe came to a standstill. To remedy this, Louis de Broglie — among other personalities such as the Italian Eduardo Amaldi and Pierre Auger and Raoul Dautry of France — proposed in 1949 the creation of a European laboratory for high-energy physics, with UNESCO's support, which encouraged the creation of scientific collaborations without a military aspect. The idea was to bring together young researchers from all over Europe, to put them in contact with each other, then to send them back to their home universities with a high level of excellence and thus restore the reputation of basic science in Europe. Eleven countries agreed on the principle, but it was still necessary to decide where the new laboratory would be located.

One group of theorists made a strong argument for Copenhagen. This is where the iconic Niels Bohr Institute of Theoretical Physics was (and still is) located, an essential place in the advent of quantum mechanics, which witnessed an intense intellectual ferment in the 1920s and 1930s. Niels Bohr himself, still alive at the time, was involved in the creation of CERN.

There were other possible choices: France, Italy... But Switzerland, with its central position in Europe and its tradition of peace, was the ideal candidate. The question of location was settled in Amsterdam in 1952, at a conference where it was decided that the Laboratory would be located in Meyrin, in the open countryside near Geneva — a decision endorsed by a referendum in 1953 by which the local population accepted the project. And the following year, 12 countries signed the CERN Convention, officially recording its birth: Belgium, Denmark, France, Germany (in fact, the FRG), Greece, Italy, Norway, the Netherlands, the United Kingdom, Sweden, Switzerland and Yugoslavia. In parallel, the theoretical section of CERN, of which Niels Bohr was a member, was initially established in Copenhagen, but this did not last long, given the distance from Switzerland.

The first objective of the collaboration was to build particle accelerators to unlock the secrets of matter. Several circular accelerators thus emerged one after the other, each larger than the one before. In the 1950s, the Proton Synchrotron was built, which, among other things, allowed the study of very strange particles that we will have the opportunity to discuss again: neutrinos. About twenty years later, the Super Proton Synchrotron was born — from then on, the PS was converted into an SPS injector, so that the protons arriving in the SPS had already acquired a certain energy, as I explained earlier.

The SPS, which sent protons against antiprotons, had its moment of glory in 1983 by allowing the discovery of the Z^0, W^+ and W^- bosons, which earned Carlo Rubbia and Simon van der Meer the Nobel Prize in 1984. These are the mediating particles of one of the four fundamental forces of nature: the weak nuclear force, involved in beta radioactivity processes. Today, the mass of the boson Z^0 is known with great precision, and for this reason it acts as a standard when calibrating the detectors.

Proton collisions are conducive to the discovery of new particles; indeed, because they have a composite structure, they can produce a wide variety of collisions, allowing a wide range of energy to be explored and thus very different particles to be observed. On the other hand, when we want to study a known particle in a narrow and well-defined energy range, electron collisions (which, to our knowledge, do not have an internal structure) are more suitable because they are "cleaner." Thus, after the discovery of the Z and W bosons, CERN decided to change its strategy by producing collisions no longer of protons, but of electrons, in order to study these new particles in detail. But when electrons circulate in an accelerator, they emit radiation called "synchrotron radiation" and lose energy because of this phenomenon. The same is true for protons, but in much smaller proportions: for the same energy, electrons lose about 10,000,000,000,000,000 times more energy than protons through synchrotron radiation. In short, it had become necessary again to build an accelerator even larger than the SPS.

This is what gave birth to the LEP, the Large Electron–Positron collider of 27 km of circumference powered by the SPS, which produced collisions between electrons and their antimaterial congeners (if I may say so), positrons. LEP made important discoveries, such as studying the decay of boson Z into other particles, in perfect accordance with the theoretical hypotheses formulated since the 1950s.

From the 1980s onwards, physicists decided to return to proton collisions, keeping the tunnel of the LEP but changing the accelerator inside. This led to the construction of the LHC, the Large Hadron Collider, in the place of the LEP. A hadron is a particle subjected to strong nuclear interaction, which ensures the cohesion of atomic nuclei, and this is the case of the components of the proton (quarks and gluons — we will come back to this later).

The LHC achieves such high energies that it recreates at the collision points the physical conditions that prevailed just after the Big Bang (in the case of collisions between lead ions). This is already remarkable, but we still hope to increase energy to highlight new particles, whether predicted or not by current theories.

CERN funding

We have arrived at Arnaud's office. You will finally have all the arguments you need to convince those around you who might wonder what CERN is for or what the point is of funding such an organization. Here it is, let me knock. I think I hear him coming...

"Hello, Arnaud, thank you for welcoming us to your office.

- Hello, both of you! Please, take a seat.

Bernard advised me to talk to you about CERN's finances... So I would like, if you don't mind, to talk first about how much CERN costs, and then explain why these investments are worth it. So let's start with the expenses... A first simple question: how much does CERN cost per year?

- CERN costs about one billion Swiss francs per year..."

A Swiss franc corresponds to a little bit less than one euro.

"... Our General Manager would say that it is the equivalent of one cappuccino per European per year..."

So I will ask you this question at the end of the visit: are you ready to offer a cappuccino to CERN every year? As for the General Manager, her name is Fabiola Gianotti.

"... This money comes from the annual contributions of the Member States, in proportion to their wealth. France, for example, contributes about 15% to the CERN budget, that is, about 150 million euros. Then, if CERN has to make major investments, it can

Fabiola Gianotti

benefit from a credit facility from the European Investment Bank. For example, for the LHC High Lumi..."

The LHC High Lumi is the High Luminosity phase of the LHC planned for 2026 to 2037, in which the number of proton collisions will be multiplied by approximately 10.

"... CERN will be able to get 250 million francs from this bank, which will then be repaid over several years. That's not all: CERN is responsible for the accelerators like the LHC, but not for the detectors, and only funds part of the ATLAS, CMS, ALICE and LHCb experiments. These experiments are independently funded by their respective international collaborations, consisting of universities, laboratories, etc. For example, the French laboratories working on ATLAS partly finance the ATLAS experiment. These laboratories or universities do not necessarily belong to Member States. Thus, the largest community of CERN users is the Americans, while the United States is not a member state of CERN! So the Americans contributed directly to the experiments, and in the case of the accelerator, they were involved in the design and production of some of the magnets. In short, when we talk about contributions, it is difficult to give a completely simple answer!

Since you're talking about the accelerator and its detectors... how much did the LHC and its four experiments cost?

- The LHC machine costs around 5 billion Swiss francs, material and personnel included. The detectors cost about 1.3 billion Swiss francs. In total, by adding the costs related to data processing, the overall cost can be estimated at 7 billion Swiss francs."

That is about 6.5 billion euros. It is said that tax fraud causes Europe to lose at least 1 trillion euros a year... so let's not aim at the wrong target!

"Finally, to conclude on the question of costs, could you tell us what CERN's energy bill is?

- All right; wait, I'll find the number in this little booklet... Here it is: if we assume that the LHC operates 270 days a year, the energy used is equivalent to the power consumed by a third of the canton of Geneva over a year, or about 1.2 terawatt hours. That's about 50 million euros — not so surprising, considering the machine you run!"

You can put it in perspective the next time you receive your electricity bill.

"Now that we've gone through the costs, it's time for THE crucial question that everyone asks... Are all these expenses worth it?

- I think there are three possible types of answers. First of all, there is the "philosophical" answer, in a sense: CERN is used to better understand the world in which we live. Humanity will never cease to ask itself questions about the functioning of the universe; its will to know is inexhaustible! In your opinion, why did the discovery of the Higgs boson have a significant impact on the public? Because people, even without perhaps fully understanding what it is, are aware that it is related both to the origin of the mass of particles and to that of the universe, so in a way to our own origin. However, to enable such discoveries, current research requires ever more efficient instruments, which can be developed only through international collaborations. This more than anything is what justifies the existence of an organization like CERN, which is international by nature..."

That is true: I personally understood at CERN the importance in particle physics of the proverb "unity is strength"!

"... Then there is the economic argument: CERN makes possible technological developments that have tangible impacts on industry. Studies have shown that for every euro invested in CERN, the return is 2 to 3 euros for contributors, particularly for France and Switzerland. New countries regularly join the organization, and they do not do so for no reason: they understand very well that, when such complex instruments are manufactured at the cutting edge of current technology, the industry concerned is enabled to become better and more competitive.

Could you give us a concrete example of this?

- Yes, there are many, but medical applications are often mentioned as a priority, because in this field everyone can understand the impact and importance that new technological developments can have. Proton therapy and some medical imaging technologies, for example, come from particle physics. Treating cancers today with particle beams is possible only because at some point physicists — at CERN and elsewhere — have learned to manufacture and operate these machines. They are now used to improve diagnosis and destroy tumours with very high accuracy.

Are there prospects for applications in industry based on technologies currently developed at CERN for particle physics?

- You know, it is always difficult to predict which technologies developed today for particle physics will find applications in industry tomorrow. Of course, we do not invent the Web every year, but history is full of examples of essential applications that were first developed through fundamental research. Things emerge as they happen, over the long term. But there are prospects, for example in the field of superconductivity..."

Superconductivity allows electrical currents to pass through certain materials (usually at very low temperatures) with zero resistance, which considerably reduces energy losses.

"... For example, superconducting cables are being developed for the LHC High Lumi: who knows but they may be used in a few years' time to transport energy with much less loss? Thanks to superconductivity, increasingly powerful magnets are also being developed, which can have applications in neurology, for example.

Indeed: I myself attended a lecture at CERN given by the Director of Neurospin, a French research centre dedicated to brain research. He said that Neurospin owes a lot to the work done at CERN, because superconducting magnets will greatly improve brain imaging. These advances could, for example, lead to a better understanding of brain-related diseases, or to a better understanding of the mechanics of thinking by studying brain activity more accurately. But let us return to the initial question about the interest of financing CERN: we have talked about knowledge in itself, and economic benefits. What would be the third argument?

- This is a strategic argument. The emergence of new technologies for particle physics or other applications is accompanied by contracts for industry. But above all, CERN is a real place of knowledge transfer, with thousands of young people coming here to do their theses in part and then returning to the business world with a high level of qualification: this has great value, even if it is not financial. It is therefore not possible to limit ourselves to purely economic spinoffs to assess returns on investment in the case of an organization such as CERN, whose training role is essential.

The last word?

- I would say that CERN is not expensive in terms of the impact it has on society. The economic benefits of the Web are certainly much greater than anything CERN will ever cost! Moreover, if the Web had not been invented by CERN, it would probably not be in the public domain and therefore free as it is today — and we would surely have to pay royalties to its inventor! In general, for all that CERN provides in terms of fundamental knowledge and technological innovation, it is a necessary organization. But I would like to add one more very important thing: CERN is one of the finest examples of international collaboration, one which shows the magnitude of the possible challenges that can be faced when men and women from all over the world come together peacefully around this noble cause that transcends nations and cultures: namely, advancing knowledge.

It's funny, you just reached almost the same conclusion as Bernard, whom we just interviewed! Thank you very much, Arnaud, for all these explanations. I think it is important to bear in mind that CERN, even if it is primarily dedicated to basic research, has reper-cussions for industry and in our daily lives, without being a financial pitfall for contrib-uting states. We will continue the visit with this in mind. See you soon!

- Goodbye, and thank you for visiting!"

Entering CERN

Well, it's time to go and find my car in the parking lot, in front of the reception area. Are you insinuating that I don't have the energy to walk? Ah, no, that's unfair. I don't know what you think of CERN in terms of size, but you'll soon realize that it's really huge. Here are three proofs of this.

First of all, at the very first meeting I attended as an intern at CERN, I naively asked where the person was located with whom we were talking by videoconference. I was expecting someone to tell me something like in the United States, Japan or another distant country. Instead, I was told: "He is on the reception side at the main entrance; he did not have the energy to come this far."

A second proof: during the visit, you will notice internal mailboxes in some corridors, called "Outgoing mail." This concerns letters for institutes outside CERN, of course, but it is also for internal mail — which still exists, even if emails have taken over. There is, therefore, an intra-CERN mail, with reusable envelopes whose age is measured by the number of scribbles.

Nevertheless, it seems to me that the best proof of the size of CERN is the agenda of activities: it would take several lives to attend them all. Here, I have on my phone the schedule of the day on Indico. Indico is the platform used at CERN to manage events, conferences, meetings, workshops, room reservations, slideshows projected during the lectures. It is a very practical system. I think that if someone who works at CERN saw the founder of Indico in a corridor, he would not hesitate to kiss him.

So, for example, if you have nothing to do at 10:30 am, you can attend *Inner detector SCT subsystem* in room 3196-R-021: good luck with finding it... But you would be wrong to deprive yourself of *Luminosity and Forward Detectors* in room 3150-R-002 at 10:30 am. You might also be interested in *Physics Standard Model / Informal WZ*, in room 40-4-C01 at 10:30 am. However, I strongly advise you to register for *Trigger & Data Acquisition & DCS*, in room 42-R-403: it's at 10:30 am. Otherwise, you will not regret going to room 40-R-D10 at 10:30 am to make sure you don't miss anything about *Core and Detector Software/ID Trigger Software*. A busy schedule, isn't it? We can count together today, there are no less than 78 working meetings — just an ordinary day. And again, this concerns only the ATLAS collaboration.

Here's my car, please, get in.

Before we start, I have something to show you in the glove compartment... It's a CERN map: you can unfold it *(see pages vi and vii)*.

There are two sites: Meyrin — where we are now — which is about 2 km long and 500 m wide, and Prévessin, which is slightly larger, although it contains fewer buildings.

In Meyrin, there are currently 460 buildings numbered from 1 to 3198. I see that you're surprised: we all were! The seemingly random distribution of building numbers is what you notice first when you come to CERN. For example, you can see that 166 is adjacent to 155, which is itself next to 187. Imagine asking a CERN expert how to get to 168 from 574, for example, the answer might be disconcerting: "Well, it's simple, you walk along building 157, in the small passage opposite left of building 263, then you join Newton Road, and there you can either turn right in front of building 2013, then turn left twice, or turn left and take Cavendish Road, the one after Joliot-Curie, which takes you directly to 168." Building 168 is sandwiched, logically, between buildings 20 and 21.

When I asked who had decided to devise such fanciful numbering and why, I was told that the numbers corresponded to both the date of construction and the type of each building. Certainly, however, this numbering does not help to get to meetings on time, at least not in the first few weeks.

I remember that once I was supposed to attend a meeting in building 222. Do you know where it is? Between 39 and 41. Did you think that building 40, towards which we are heading, would be between 39 and 41? No, come on, forget that crazy idea! On the other hand, remember that if you have an appointment at 222, you will have to pass between 39 and 41, and *certainly not* between 221 and 223, even if they are close to each other. You smile, but another time I had an appointment in number 30, and I was near 31; so, without looking at the map, I thought I wouldn't have much distance to walk. An elementary mistake that cost me a long epic run in the rain!

Fortunately, if you feel lost, there is a miracle solution for finding your way: the CERN-map app! You might suppose in retrospect that people deliberately forgot their arithmetic and assigned the numbers without any logic in order to enjoy coding this application afterwards.

All right, no more jokes, let's start. We're going to come out of the car park to the main entrance. Wait a minute... Did you see that? The barrier always takes a little time to get up. That's because there's a little camera there that takes a picture of our licence plate. No one laughs about security here, I can assure you of that! In fact,

The security post at CERN's entrance

you will notice that each car has its trunk searched at the exit. In case we steal a particle detector...

We arrive here at the main CERN's entrance. This main road, which we will take, is the Pauli Road, and immediately to the right, you can see a big and beautiful road: the Einstein Road.

Einstein Road

Here, all roads, alleys and squares are named after physicists, which is both consistent with the place and... unusual! And I have a theory: we can quantify the fame and importance of a physicist according to the quality of the road, alley or square to which he gives his name. I'll stop for a moment (there's no one behind us). This superb Einstein Road, whose end can hardly be seen (it is indeed very long), is wide, the concrete is smooth, and the cars are well parked. Translation: Einstein is a leading physicist. Now look to the left: the Yukawa Road is narrow, it doesn't make you dream and — worse — it's a dead end.

This is quite surprising and unfair to Yukawa, who, although not well known by the general public, is constantly quoted in works about the Higgs boson, because he gave his name to the constants characterizing the intensity of the coupling between matter particles and the Higgs boson. He was also the first to theorize the strong interaction. But, because of the star system, with an unknown name, you have a road difficult to drive on... Let's continue our way, slowly: on the right is the magnificent Democritus Road, almost the twin of Einstein's. And on the left is Lawrence Road, which is pointless and leads to nowhere. Likewise, who cares about Scherrer Road? What about Greinacher Road? Who even knows that these physicists existed? I mean, that's life...

Now that we are at the heart of CERN, we will get to the heart of (the) matter. We will begin the real visit, the one that follows the path of the protons. Where does this journey start? In the hydrogen bottle, for sure. Do you remember the name of the first accelerator into which the protons are injected? A hint: this accelerator is linear... This is LINAC 2, and that's exactly where we're going.

We now follow Salam Road, and we will turn right onto Weisskopf Road. We are not very far away — we just have to fork on Faraday Road... Let's stop here, that's fine. We will enter this large hangar opposite; it is building 150; we don't have much distance to go to reach LINAC 2... But let's start at the beginning and take a look at the hydrogen bottle, which is the real starting point for all the protons that circulate in the various accelerators. After you.

Yukawa Road

LINAC 2 and LEIR

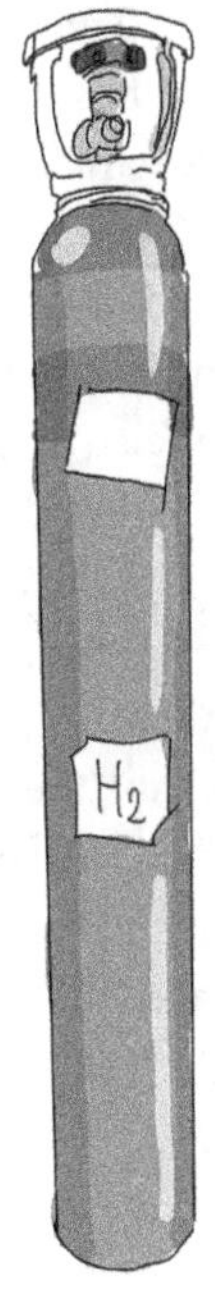

There's the hydrogen bottle! Well... At the risk of disappointing you, this bottle is fake — the real bottle, behind it, is not visible. But it doesn't matter, knowing that it is there is already a great source of satisfaction. Are you wondering whether it is necessary to replace this bottle often? During the operating period, only about 2 nanograms of hydrogen are accelerated per day. It would therefore take more than a million years to accelerate one gram of hydrogen, and the bottle contains several hundred of them... However, it must be replaced every year, because much more gas is consumed than is accelerated.

I say "hydrogen," but in reality, it is protons that are accelerated. Between the bottle and LINAC 2, it is necessary to extract the electrons from the hydrogen atoms, in order to keep only the protons. The principle is simple: when the hydrogen gas comes out of the bottle, it is placed in a chamber where it is so hot that the protons separate from their electrons to form a soup that is called, in more scientific language, a plasma. This is then ejected from the chamber at full speed and subjected to an electric field, so that the protons, with a positive electric charge, continue to travel in the LINAC 2, while the negatively charged electrons go backwards.

Our protons are therefore found at the entrance of LINAC 2! The acceleration phase can begin. But what is needed for acceleration? First, an electric field. Here it is of sinusoidal type, which means that the proton alternately crosses acceleration and deceleration zones during its propagation. But since we do not want to slow down the proton, in the deceleration zones we must hide the electric field from it. To do this, the trick is to use large copper tubes through which protons are subjected only to the electric field where it accelerates them: then they gain energy, and therefore speed up. But that's not all! These tubes also contain small magnets that focus the proton beam, so that it remains as thin as possible along its path.

Finally, I must stress here an important function of the LINAC: it begins to group protons in *packets* and sends them together to the next accelerator, the Booster. Make a note of this — it's important — in all accelerators after LINAC, protons circulate in packets! We will have the opportunity to discuss this later in more detail with a specialist.

Follow me and look through the glass of that door. Here it is!

We can't go any further, but we are in the front row to admire this beautiful linear accelerator, whose colour is reminiscent of David Bowie's orange haircut embodying his character Ziggy Stardust. (I know... Everyone has their own references.)

On leaving LINAC 2, as I told you in the Globe, protons are injected into the Booster, then the PS, etc.; but they are not the only ones... Lead ions can also be accelerated. It is not LINAC 2 that does it, but LINAC 3. Let's gain a little height, in the large hangar 150 adjacent to building 363, where we are.

Let's take the stairs in front of us, which lead to a footbridge. Here it is. From this platform, we are also very well placed to observe below... The LEIR, the *Low Energy Ion Ring!*

LINAC 3, which we will not visit because it is similar to LINAC 2, accelerates lead ions and injects them into the LEIR at the white wall, which you see at the bottom, in an anticlockwise direction. The LEIR accumulates the LINAC 3 beam, increases its density, then accelerates it and sends it to the PS. The ions then continue to travel through the PS, then through the SPS, then through the LHC, until they collide within the detectors.

Finally, thanks to this complex of accelerators, we can achieve collisions of protons against protons, lead ions against lead ions, and, more rarely, protons against lead ions. But I told you that our visit would follow the path of the protons, so let's go back to them.

After being accelerated in the LINAC 2, the Booster and the PS, some protons go to the next accelerator: the SPS. But some others are directed to... the antimatter factory, which we will visit right away! Please, get in the car; the journey will not be very long. So... Rutherford Road — no, it's not that one... Oppenheimer Road... Yes, we are here. It's this big building; I'll park.

The antimatter factory

You know we've lost something. Keep calm, it's not our keys or our wallets. But all of us, and the universe, have lost antimatter — an unfortunate episode that happened 13.7 billion years ago.

It is assumed that at the time of the early universe, there was as much matter as antimatter. Then, antimatter disappeared after a homeric battle with matter. The tiny portion of matter that survived this carnage is the one we know today. In fact, you

The antimatter factory

and I, like the ordinary matter of the visible universe, are remnants of the greatest war in the history of the universe, which probably lasted only a fraction of a second.

Why did matter win at the expense of antimatter? Very good question! If you have a convincing answer, make yourself known to the Swedish committee that awards the Nobel Prize. However, at CERN, several experiments are underway in an attempt to answer this question. In particular LHCb, one of the four large LHC detectors, is trying to understand the origin of this asymmetry between matter and antimatter. There is also hope for a response from the ALICE detector, since the collisions of heavy ions in the LHC are so energetic that they reproduce the physical conditions that prevailed at the time of the primordial universe. If, like me, you bitterly regret that there is no *Back to the Future 4*, you can console yourself with the knowledge that the LHC, under your feet, is by far the best time machine ever. Here, we are going back not to 30 years ago, but to 13.7 billion years ago!

We arrive at the entrance. I agree with you, this panel can be a bit chilling.

But don't worry, the radiation we receive here is ridiculously small. So you can follow me inside. Just be careful to avoid antimatter — if it squirts towards you, it could annihilate you. Do not forget: one you + one anti-you = no more you at all, with release of a large amount of energy

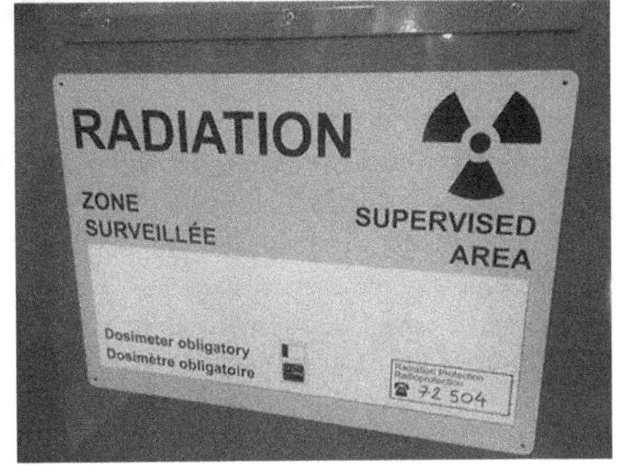

in luminous form. That being said, as far as I am concerned, I would be happy to meet my anti-me, out of simple intellectual curiosity.

This large hangar is unique, in that it is probably the place in the universe that contains the most antimatter! Looking down from the footbridge where we are, we can contemplate an antiproton decelerator, which has a pretty little name: ELENA, otherwise Extra Low ENergy Antiproton. Indeed, here, we don't try to make collisions, but, on the contrary, to trap antimatter: here in this case, antiprotons. If you want to study something, it's better if it doesn't move in all directions; so here, we slow down antiprotons by giving them as little energy as possible.

But how do we create antimatter? First, the PS — as a reminder, the Proton Synchrotron — sends some of its protons to a metal target, resulting in the production of many particles possessing a very wide range of energy. However, among these are some antiprotons that have just the right energy to be injected into the Antiproton Decelerator, called AD. ELENA, just

The ELENA decelerator

below, is another decelerator connected to the AD, which is even more efficient. Obviously, all these decelerators operate within a vacuum and direct the antiprotons so that they do not meet any piece of matter!

What do we do with these antiprotons next? Well, we study them. You will often read this definition of antimatter: particles that have the same properties as their material counterparts, except that some of the quantum numbers that characterize them are of opposite values, such as their electrical charge. For example, this is +1 for a proton, and −1 for an antiproton. But is this charge really exactly the opposite? And are these "same" properties really the same? Several experiments here are therefore trying to measure the charge, mass, etc. of antiparticles with the greatest possible accuracy.

For example, one of the experiments, ALPHA — which in 2011 managed to trap anti-matter atoms for more than 16 minutes, a record! — has already shown the neutrality of the antihydrogen atom, consisting of a negatively charged antiproton and a positively charged positron (or anti-electron). This same experiment also concerns the spectrum of the antihydrogen atom, i.e., the light emitted by the positron when it loses energy. At the end of 2016, ALPHA succeeded in observing a line in the spectrum of the antihydrogen atom. However, this line does not seem to present any difference from its cousin, that of the spectrum of the hydrogen atom which corresponds to the same loss of energy, no longer of the positron, but of the traditional electron. This result is in perfect agreement with the reference theory in particle physics, the Standard Model, which is both very satisfying and... frustrating: physicists expect that some experiment will contradict the theory to open the way to a new physics, and therefore to a finer understanding of our universe.

But that's not all. Let's move a little to the left, further along the footbridge... Do you see the large square panel "GBAR" on the wall below?

This is an outstanding experiment. GBAR (Gravitational Behaviour of Antihydrogen at Rest) will soon use the new performance of ELENA to find out how an antiproton behaves in a field of gravitation: does it fall? Does it go up? Does it levitate? Let's say it falls: at what speed? If I drop a metal ball and an antiproton from the same

The GBAR experiment

height and at the same time, do they reach the ground simultaneously? To be continued...

UF37I see the little smile appearing on your face: don't tell me that all this is not very useful! Antimatter is already used in medical imaging, and it is hoped that antiprotons can be used to annihilate cancer cells much more effectively than simple protons. And if you still have any worries, I would like to make them disappear with these anti-worries. You should know that it would take the age of the universe to make enough antimatter here to develop an antimatter bomb. At most, in terms of energy, the antimatter so far produced at CERN could heat a coffee cup at 30°C. So, don't panic, we're not going to blow up the Vatican tomorrow with such a bomb, as in the movie *Angels & Demons* with Tom Hanks.

On this last reassuring note, I propose to close this antimaterial episode, this virtual anti-episode, and go back to the car. One bit of good news, which we will not forget to highlight, is that we are still whole!

Well, let's get back to our protons. We followed those that go from the PS to the antimatter factory; now we will follow others, those that are injected into the SPS, then into the LHC, before colliding. In the LHC, we accelerate protons as much as possible to make them acquire the greatest possible energy. But it is not enough to accelerate protons: they must also be guided, so as to follow the circular trajectory of the LHC. But how do we guide protons? Through what

Inside the cryomagnets test facility

mechanisms and applying what principles? To give you an idea, there's nothing like going to SM 18, a large facility where all the magnets currently in service in the LHC tunnel were tested.

SM 18 is located a little outside the Meyrin site where we still are. When we are back at the Globe, we will take the road to the left, along CERN.

We cross the customs post. I have never seen a single customs officer. Times have changed a lot! I was told that before the creation of the Schengen area, Swiss people who came to France to shop received tickets that allowed them to recross the border from France to Switzerland with a certain quota of goods. In that way, the Swiss prevented the Swiss from buying too much food in France, where prices were (and still are) much lower. We are far from such restrictions now.

It's somewhere on the right... Here! We have just arrived at SM 18. This is where the LHC magnets were tested after being assembled in a nearby hangar. Let's take a look around.

In SM 18, the right question to ask is: how do we guide and accelerate particles? The recipe is relatively simple, at least in theory. You actually need only three ingredients: a good spoonful of electric field, a few strands of magnetic field, and a delicate sprinkling of vacuum overall. Electric fields accelerate protons, magnetic fields guide them, and the vacuum prevents them from interacting with matter and being slowed down or deflected.

As I have already pointed out, with respect to collisions, half of the protons injected into the LHC from the SPS circulate in one direction and the other half in the opposite direction. But how do protons travel simultaneously in both directions in the accelerator? Come closer...

We are in front of a magnet whose inside can be seen. Do you see the two large pipes that come out at half height? The two beams of protons circulate through them: one pipe is reserved for clockwise motion and the other for counterclockwise.

This magnet is called a dipole: it is the most common in the LHC; there are 1232 of them in the tunnel. Each one is 15 m long and weighs 35 tons. Dipole magnets are used to guide the proton packets so that

Dipole magnet as seen in cross-section

they follow the trajectory of the tunnel along their two pipes. Protons circle the LHC more than 11,000 times per second, and thus cross the border more than... 2,640,000 times per minute, since the LHC meets the border four times. In other words, for them and for customs officers, it is a good thing that France and Switzerland are part of the Schengen area, so there are fewer formalities.

Since protons move very fast, because they have a large amount of energy, magnets must generate very, very, very strong magnetic fields to deflect them (I say "very, very, very," because the exact figure of 8.3 teslas may not mean much to you). However, to obtain such magnetic fields, you need... a cold environment. It must be colder than interstellar space, if that is more concrete to you than 1.9 K, or −271.3°C. In short, the LHC is the largest freezer in the world. This temperature is reached in the central block of the magnet thanks to helium, which circulates in this tube that you see in the middle at the top: this is what gives the magnets their superconducting properties. Superconducting means that the electrical resistance in the magnets is strictly zero: it allows huge currents of about 12,000 amperes to pass through them, currents which are at the origin of very powerful magnetic fields. This figure of 12,000 amperes may not mean much to you either — in fact, it's so high that it probably doesn't mean anything to anyone — but when you know that a conventional office light bulb works at 0.1 amperes, you can only be advised against putting your fingers in the magnet.

In addition to these 1232 dipoles, 392 magnets called quadrupoles are arranged, which ensure the cohesion of the packets by refocusing the beam regularly. Can you imagine? To install all the magnets in the LHC, as they are very fragile, they had to be transported over a cumulative distance of 30,000 km at a speed of... 2 km/h. I think we can think of the drivers who had to drive so many kilometres in a tunnel where 3 km/h is already exceeding the speed limit.

As for the vacuum, here, too, the LHC is remarkable: 10^{-10} mbar. "Empty" is not really an appropriate term here because the vacuum of the LHC, although very high, is not perfect; there are still residues, in this case about 3 million molecules per cubic centimetre.

It's like the bones in the red mullet: no matter how many of them you remove, there are still some left. Nevertheless, there are fewer particles in the pipes through which protons pass than in the lunar atmosphere. On their own tiny scale, protons must feel very lonely: one is almost moved to think of it.

With the two main types of magnets, dipole and quadrupole, as well as others that make even finer corrections, the protons grouped in packets remain well together and follow a circular trajectory: very well. But at some point we must accelerate them, or more precisely, give them more and more energy. This is done by means of electric fields in accelerator cavities arranged in a straight line at a specific point in the tunnel (near the CMS detector), through which protons pass more than 11,000 times per second. And when the protons overflow with energy, they are directed towards the detectors and against their companions — who arrive fully inflated in the opposite direction — in order to produce collisions.

Now, let's hurry back to the car, because another meeting is coming up…

To complete what I just told you about particle acceleration, there is nothing like a little meeting with Elias Métral, who is waiting for us in building 10, not far from the LINAC 2 we visited earlier. Elias is a specialist in what we call beam instabilities, that is… all the problems I haven't told you about! I think that until we talk to Elias, we'll have no idea of all the challenges we have to overcome in order to create a beautiful beam of protons at very high energy.

We are back in Meyrin. Einstein Road, Yukawa Road: you should feel like you're on familiar ground now. But this time, we will not follow Faraday Road, but Becquerel Road. All right, let's stop right next to the entrance to building 10.

Elias works on the firstfloor of this building. I have no doubt he will have his usual enthusiasm! Let's take this staircase to the left… I think his door is ajar. Yes, it is this office.

Interview with Elias Métral

Beam instabilities

"Hello Elias! I see you were waiting for us.

- Hello, both of you! You just got back from SM 18, right?.

Absolutely, a good guess.

- So you were able see some of the magnets that we'll talk about again.

Indeed! But first of all, could you explain to us exactly in what field you work?

- I work in the BE (Beams) department, which deals with everything related to particle beams, from accelerator physics to collisions for the LHC or experimental areas for

other accelerators, such as the antimatter factory. This department is divided into several groups: I am part of the one called ABP, Accelerators and Beam Physics, which deals with theory and simulations for particle beam dynamics. In this group, I lead the HSC section, for Hadron Synchrotron Collective effects, which focuses on beam instability effects.

We are going to talk about these instability effects, of which you are a specialist, but first, following the visit to the SM 18 hangar, we would like to know a little more about how the accelerators work.

- Of course! Do you know the different synchrotron accelerators at CERN?

Yes, but.... What is a synchrotron?

- It is a type of particle accelerator, circular, in which — by definition — particles always follow the same trajectory. At CERN, there are, among others, the Booster, PS, SPS and LHC, which are based on the same principles but each have their own specificities."

I believe that the drawing with the various accelerators will be useful again.

"...Since we look for rare particles, such as the Higgs boson, we need to have high luminosity to produce a maximum of collisions at the highest possible energy..."

The luminosity is the number of collisions that can occur per unit area over a period of time.

"... Depending on what we want to study and the performance of the detectors, particle physicists ask us — as accelerator physicists — to provide a specific luminosity. This is achieved by minimizing the size of the beam at the collision points, maximizing the number of proton packets circulating in the accelerators, maximizing the number of protons in each packet, and propelling them to the highest possible energy.

But how are these packets of protons formed?

- First, at the output of LINAC 2, protons are gathered in large packets, which are accelerated in the Booster from 50 MeV to 1.4 GeV..."

These are energy units: 1 GeV, i.e., 1 giga-electronvolt, is equivalent to 1,000 MeV, i.e., 1,000 mega-electronvolts.

"... Then 6 large packets are sent into the PS, which not only accelerates them to 26 GeV, but also cuts them into 12 smaller packets.

So we get 72 small packets?

- Absolutely. The PS sends them to the SPS, and this process is repeated 4 times: there are then 288 packets in the SPS. The SPS accelerates them to 450 GeV and then sends them to the LHC. After about 20 minutes, about 3000 packets circulate in the LHC in both directions.

Do we accelerate the packets at the same time as we inject them?

- No, no! From one accelerator to another, the packets are first injected, then accelerated by giving them an energy gain each time they pass through accelerating cavities. During this phase, the energy of the proton packets is increased step by step and synchronized for all packets.

And how do we keep these packets compact? After all, at very high speed, a bunch of protons could dissociate very quickly!

- For that reason, in the accelerator, particles circulate in vacuum chambers like this one.

Ideally, the particles should circulate exactly in the middle of these chambers. But in each

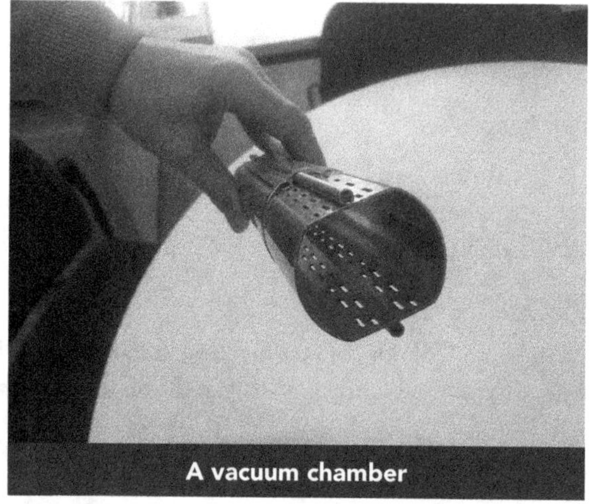

A vacuum chamber

packet, there are about 100 billion protons — about as many as neurons in a developed human brain..."

That is an average, of course...

"... These protons each occupy a certain space, so that not all particles can follow the ideal trajectory in the centre. With respect to the axis, some protons are shifted up, down, or to the sides. To ensure that the protons do not deviate too much from the ideal trajectory and that the packets do not break up, horizontal quadrupole magnets are used, which bring the protons that move to the left or right back to the centre, as well as vertical quadrupole magnets, which bring the protons that move up or down back to the centre."

Remember: at SM 18, I told you about these quadrupole magnets, which, alternatively with the dipole magnets, focus the beam and guide it.

"Now let's talk about your area of expertise... What is a collective effect?

- This is an undesirable effect linked precisely to the large number of protons per packet and the large number of packets, which can cause beam instabilities and degrade its quality.

Are there many of these effects?

- Yes, but here we mainly study four of them.

So, to keep up the suspense, tell us about them, from the least embarrassing to the most embarrassing.

- Okay, but this hierarchy depends on the accelerator: let's take the case of the LHC. The first effect is the electrical repulsion between protons, which are of the same charge and accordingly repel each other. Therefore, you can't reduce the size of the packets indefinitely. This is called the space charge effect, which is most noticeable at low energy because, when they move fast, protons each generate an attractive magnetic force which counterbalances this repulsion force. So in the LHC, where protons are already at very high energy, this is not the predominant effect.

But it would be more so in the Booster, PS and SPS. Second effect?

- You know that, in nature, opposites attract each other: so the protons of the beams, which circulate in the middle of the vacuum chamber and which are positively charged, tend to attract electrons of negative charge to the surface of the walls of the vacuum chamber. These electrons thus create small disruptive electromagnetic fields...

That we can correct?

- Yes, but not completely. Ideally, we would have chambers with zero resistivity and always of the same shape, which is impossible. In practice, the chambers are made of stainless steel with a copper coating inside to increase conductivity; despite this, these disturbing fields remain important. That is why 84 eight-pole magnets are used to stabilize the beam.

All right! What about the third effect now?

- Within the detectors, the beams that travel in opposite directions meet to produce collisions. But, among the 100 billion protons in each packet, only about 20 really collide! The others see and repel each other. So, as they cross each other, the packets become larger and larger, so less and less dense. Their encounters then produce fewer and fewer collisions, and when there are not enough collisions, after 10 or 20 hours, the beam is dumped."

In other words, into the trash can!

"So we come to the fourth effect...

- I kept the suspense going, didn't I? The fourth effect is all the more harmful because it was not known before the LHC was built. This is the "electron cloud" mechanism. For a while now, I've been talking to you about empty chambers, but in reality they are never really empty: in particular, there are always residual electrons lying around..."

Remember the red mullet...

"... For example, an electron almost at rest will be accelerated by the passage of a packet of protons, and will bounce off a wall of the chamber and leave this wall with one or more other electrons. Then, as the next packet of protons passes, these electrons will bounce off the opposite wall again and carry other electrons with them. Through an avalanche mechanism, all these electrons will form a large cloud. Protons travelling at very high speeds see these electrons, which create large disturbances in their movement. Wait, it's already... Oh, I have a meeting coming up. Did you have any other questions?

One last, quickly: in the face of all these problems, what does your daily life look like?

- As head of section, my priority is to deal with the team, whose work is divided into three parts: theory, simulation and measurements in accelerators, when we are given time to play with them. This allows us to test our models to understand our accelerators better and propose solutions to boost performance. At the same time, I am co-director of

JUAS, a school dedicated to accelerator physics in Archamps, France, about 30 minutes from here. Sorry, I have to leave...

All right, thank you for telling us about all these collective effects, and... have a good meeting!

-See you soon!"

I think we have just had a fine overview of some of the difficulties encountered in particle acceleration. I hope that this conversation has made you realize that accelerators, and in particular the LHC, are a real technological achievement. Now, if someone tells you about particle acceleration, you will hear beneath this expression "a million words," to use Philaminte's hyperbole in *The Wise Women* by Molière. And you are still far from being at the end of your surprises!

Following this conversation with Elias is a very good time, I think, to visit the CCC, the CERN Control Centre. This is the main control room of CERN, dedicated to the various accelerators. Let's get in the car for new adventures...

As I told you earlier, CERN is divided into two sites: the one in Meyrin where we are now, and the one in Prévessin 6 km away in France, where the CCC is located. The Meyrin site is unique, in that it crosses the Franco-Swiss border, but we do not know exactly where. In any case, it is certain is that, while we entered CERN through Switzerland, we will leave it through France.

Now, look at this imposing structure... Can you guess what it is? A control tower, like those at airports? Game over. In fact, when you see the size of the control room, you will realize that it could not fit in this tower. A surveillance post? Game over again. It's actually... a water tower. Yes, I was also quite surprised when I was told; we would not necessarily have expected a water tower to be the highest point of CERN.

We approach the French side entrance: the Charles de Gaulle entrance. It seems that, when visiting CERN, the General said he would enter only through France, and that this entrance was created especially for him.

I hate this dangerous roundabout; I'm always afraid to run over a cyclist. Do you see the Y bus a few cars ahead of us? It is the only bus that goes to CERN and passes through the small towns in the area. I can guarantee you that for many people who work at CERN and live on the French side, this bus is a blessing.

I see the entrance to the Prévessin site, you can already take out your visitor's pass. The security conditions are the same as in Meyrin.

Welcome to Prévessin! I'm going to park in the parking lot just to the left. We have arrived in front of the CCC building, the CERN Control Centre.

Interview with Django Manglunki

CERN Control Centre

The CERN Control Centre building

The entrance is a little further, follow me. We will have the opportunity to see these large windows from the inside. Before going through the automatic door, don't miss the beautiful CCC logo on the side. Ah, here's Django.

"Hello, Django! Thank you very much for welcoming us to the CCC and for sharing your experience with us.

- It's my pleasure! Come with me; the control room is at the end of this restricted access corridor on the left."

Panoramic view of the CCC interior

I'm starting to see the screens... I've never been there before. I'm discovering this place at the same time as you. Wow!

"But Django, it's huge!

- The room is 625 m^2; it has the shape of a square of 25 m per side.

You were project manager of the working group responsible for making this control room... Could you tell us a little bit about its history?

- Yes; this room has existed for eleven years. At the time of its construction, the goal was to make it as comfortable as possible, in order to optimize the efficiency of the work, day and night. Specialists were hired, who focused on acoustics. To be able to chat in a corner without being too disturbed by the surrounding noise, we put in large separate windows (instead of a bay window), so that the sound doesn't reverberate, besides carpet on the floor, acoustic tiles on the ceiling, and absorbers on the walls and behind the consoles.

Has this whole system proved its worth?

- It was tested on the day of the inauguration when there were 200 people and, indeed, the sound was well muffled."

There are computer and TV screens everywhere. It's impressive! Doesn't it remind you of Wall Street? Except that here, we don't deal with stocks, but essentially with protons.

"How many screens are there in this room?

- There must be something like 400 screens, connected to 120 classic computers — not to mention about fifty screens hanging on the walls.

It's incredible. But what is the point of having so many large screens on the walls?

- Each screen has its own specific purpose, so they are all useful! They show real-time information, such as current intensities, the number of particles in each beam, the states of the cooling elements, the values of the magnetic field corresponding to the energy of the protons...

Indeed, it seems that each one shows distinct information. The room seems to be divided into four sections, doesn't it?

- Absolutely — four "islets" actually, which correspond to the four former control rooms. With the arrival of the LHC, we understood the importance of such a common room: since the beam properties are transmitted from one accelerator to another, it is useful to gather the information from all the accelerators in the same place."

In fact, if there is a problem at the beginning of the accelerator chain, it will have consequences for the collisions at the other end of the chain.

"Could you describe the role of each of these four islets?

- Of course; let's place ourselves in the middle of the room.

The islet in front of us on the right is dedicated to infrastructure and cryogenics, therefore to cooling. The other islets are dedicated to the different accelerators; the one in front of us on the left represents all the accelerators on the surface, such as the Booster, the LINAC 2 and the PS. Now, let's turn around...

... The islet on our right is that of the SPS, which is located underground, as well as the LHC, whose islet is on our left.

Islet dedicated to accelerators on the surface

Islet dedicated to infrastructure and cryogenics

Islet dedicated to the LHC

Islet dedicated to the SPS

How many people are in this control room?

- At least 9, day and night: 2 for the cryogenic and infrastructure part, 3 for the PS complex, 2 for the SPS, and 2 for the LHC.

Let's take the LHC, for example... What should the people working here watch out for?

- First of all, people here do not only monitor! For the LHC, for example, it is necessary during operation successively to prepare the injection of the beams, send pilot beams to ensure that the machine is properly adjusted, make all necessary measurements to apply any corrections, make the injections, put the machine in the acceleration phase, and then start the collision phase, which lasts about ten hours. It is therefore mandatory to be active in order to operate the machine in good conditions!

And in case of trouble, is there an emergency stop button for the LHC?

- Uh, yes.... That's this red switch."

Can you imagine: if the person who works there spills his coffee, and with a wrong move inadvertently changes the switch from the

A dangerous cup of coffee...

"BEAM PERMIT" position to the "BEAM DUMP" position? Not very reassuring, this cup.

"And this kind of panel, the big grey one with buttons and lights everywhere, what is its function? It's like facing the dashboard of a spaceship, as in Star Trek...

- Yes, it reminds one a bit of science fiction of the 1960s. It is in fact a safety device that manages the inputs and outputs in the accelerators. We want to make sure that there is nobody in the danger zones when the beam is circulating. In case of emergency, the beam can be prevented from reaching these areas, using obstacles or dumps. In addition, until everyone has brought back the access keys, so that we are sure everyone has left the risky areas, we cannot restart the machines."

So it's better to avoid taking the key home accidentally...

"Oh, there is a beautiful row of champagne bottles under the screens!

- It is a tradition: each bottle represents a milestone event to celebrate, like the first time the

A control panel

machine was turned on, the first beam injection... They are not only reserved for events that deserve Nobel Prizes — otherwise we would not drink much.

A last personal question, to conclude: what does your work consist of, on a daily basis?

- One week out of five or six, I am SPS coordinator: with my team, during that week, I am responsible for the smooth functioning of the SPS. We relieve each other 24 hours a day, with slots of three times 8 hours, and thanks to the Web, we can even verify remotely that everything is going well without having to call people on site.

And the rest of the time?

- I am in charge of the LEIR machine, and I prepare the beam periods with heavy ions. So most of the time, I work on LEIR. Soon, we will accelerate xenon nuclei; we have never done that before.

So there are new collisions in perspective! Thank you very much, Django, for this very informative visit of this beautiful control room.

- My pleasure."

The champagne bottles

Django Manglunki

And now we come to collisions, so to the detectors! Considering that I spent five months in the ATLAS communication group, this is the detector that I know best and would like you to discover. But since I'm keeping the original for the end of our tour, I will present you with a model of it, which is located in building 40. This is on the Meyrin site, let's take the car back without further delay.

I will take the opportunity of the return trip to briefly expose the three main research themes of ATLAS. That way, when I talk to you shortly about the general operation of the detector, you will already have an idea of why so much effort has been put into it.

First of all, ATLAS — as well as its alter ego CMS — is actively involved in the study of the Higgs boson, which was independently observed in 2012 in both experiments. Of course, we now know that this particle exists but, as Bernard told us earlier, the Higgs boson is a story that is far from complete! We are entering a phase of taking precision measurements, in order to better understand the properties of this boson, and in particular its couplings with the other particles.

Then, other studies are aimed at testing the validity of what is called the "Standard Model," which is the bedrock of particle physics, and which I will present to you later. The challenge is, for example, to measure the masses of two specific particles, the top quark and the W boson, with the greatest precision possible and to compare them with their masses as predicted by the Standard Model.

Finally, the ATLAS experiment looks for clues to what is called "new physics." Indeed, science is confronted with mysteries that the Standard Model of particle physics, although very elaborate and robust in its predictions, fails to answer questions such as: what is dark matter? What is dark energy? Where does the asymmetry between matter and antimatter come from? To hope to answer these questions, it is imperative to extend this Standard Model or develop a more global theory, which is referred to as "new physics." For this purpose, ATLAS, as well as the other detectors, is on the lookout for a new particle or an unknown phenomenon.

Among the possibilities for extending the Standard Model is the Supersymmetry theory, which provides a so-called supersymmetric partner for each elementary particle. For example, the electron would be associated with the selectron, the gluon with the gluino... Despite the enthusiasm attracted by this theory, no trace of Supersymmetry has yet been detected.

We have just passed building 221, so we are not far from building 40. I can already see Marie Curie Road. I'll park here.

Building 39 is (with 38 and 41) one of CERN's hotel buildings. And just behind it is the famous building 40. Pretty original, isn't it? It contains about 300 offices, mainly occupied by ATLAS and CMS physicists, as well as four amphitheatres and numerous meeting rooms.

Building 40

Bike station next to building 40

Many bicycles are parked in front of the building. We are environmentally friendly at CERN! There is a free bicycle loan service. For those who live in the region, this is great advantage, especially in summer. I also want to show you, right next to the bike station, this sign, which shows the name of the driveway where we are. You may know that one of the most important physicists of the 20th century was Wolfgang Pauli, famous among other things for having predicted in 1930 the existence of a very strange particle, the neutrino, whose experimental proof dates back to 1956. We will have the occasion to talk about the neutrino again, but if I refer to Wolfgang Pauli now, it is because the name of the driveway on the sign is: "Wolfgang Paul," in white letters on a blue background.

You can imagine my first reaction to this sign, which I now see every day: "But where is the 'i'?" I obviously rushed onto the Web to learn about this gentleman, and I discovered that Wolfgang Paul received the Nobel Prize in Physics

Sign next to the bike station

The statue of Shiva

in 1989 "for the development of ion capture." It is time to pay tribute to him, because, as you will agree, bearing such a name is frankly unfortunate. It would be like being called Pablo Picass!

Before getting inside, I would like to show you one last curiosity: the statue of Shiva, between building 40 and the CERN hotel. It was offered by India about ten years ago. Shiva's dance, which gives life to the Universe, is associated here with the dance of elementary particles. It's good to know, if someone asks you why she's there.

Let's enter building 40 through the main entrance. The atypical architecture testifies to the fact that this building is the VIP building, *the place-to-be*! We are in the upscale districts of CERN. The first time I came here, I thought to myself: well, it's a pretty nice building. But I didn't really understand my privilege until the day my colleagues, and even my roommates, said to me, "What? You work in building 40? The one with the cafeteria in the middle? You're so lucky!!" Don't miss, just on your right, the beautiful model of the ATLAS detector.

The model of the ATLAS detector

In reality, what is its diameter, in your opinion? The correct answer is about 25 m. And its length? 46 m. Yes, you have to use such monsters to observe such small particles! As you can see, the detector has cylindrical symmetry: I will explain what happens inside, from the centre, where the collisions occur, to the periphery of the detector.

In the axis of the detector, in the centre, the two proton beams travelling in opposite directions meet almost at the speed of light (i.e., about 24 million times faster than Usain Bolt at the peak of his speed). First, let me point out that I will talk to you about the particles as if they looked like beads. This image is in fact completely false, because of what is called in quantum mechanics the wave–particle duality. But this intentional distortion will help to make things more understandable.

The majority of protons pass by each other without interacting, but sometimes some of them meet frontally, and BAM! They collide and generate a myriad of new particles with more or less long lifetimes. The particles produced can be of all kinds; the only condition is that the total energy must be conserved, in other words that the energy of the two protons that have collided is equal to the sum of the energies of all the particles resulting from this collision.

I would like to stress here a point that we will discuss later, but that we should bear in mind now: these new particles did *not exist before the collision* — they were born from the energy generated by the collision. When you bring an egg into contact with a pan, the yolk that comes out of it was already in the egg before the collision. At CERN, it's completely different: it's as if the energy released by the collision of the egg against the pan was converted into cherries or raspberries — in short, into something that didn't exist in the egg before the collision!

Thanks to the detector, we will be able to study the particles resulting from the collision. What properties of a particle can be known? There are three answers: its electrical charge, its energy and its momentum, a quantity that is expressed in terms of its mass and speed. This is why the detector is divided into several sub-detectors: each one provides specific information to measure these different properties, by which the particle can finally be identified. There are generally three types of sub-detectors: pixel detectors, calorimeters and muon detectors.

First, the pixel detector (located within a radius of about 1 m around the beam) makes it possible to reconstruct the trajectory of the charged particles from the collision. Imagine that a charged particle is the Little Thumb. When Little Thumb walks around in the pixel detector, he deposits small stones in different places in his path. The pixel detector is immersed in a very intense magnetic field, and because it is charged, Little Thumb is sensitive to it: this has the effect of bending its trajectory. This curvature is all the more important because Little Thumb is slow and light. Then, the game is to find the path taken by Little Thumb from the pebbles he has left behind him. This pixel detector is a real technological gem. Imagine: the part closest to the collisions alone contains about 80 million reading channels, which make it possible to track Little Thumb with a resolution of only a few micrometers!

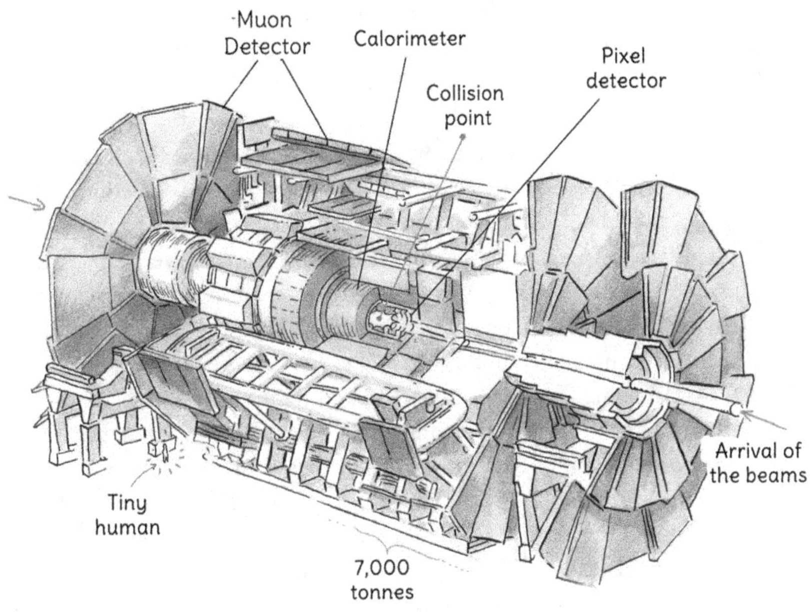

Diagram of the ATLAS detector

Behind the pixel detector is the solenoid magnet that creates the magnetic field I have just mentioned, and behind it is the calorimeter, with a radius of about 3 m. And the calorimeter is the Ogre! Why? Because this is where some particles deposit all their energy. Indeed, in the calorimeter, a particle will interact with a dense substance and thus produce a sheaf of other particles — such as electrons and photons — between which its initial energy will be distributed. Each sheaf is characteristic of the particle that gave rise to it: for example, the larger the sheaf, the more energy the original particle had at the beginning. It's a bit like fireworks: you observe the shape and intensity of a sheaf, and you deduce the properties of the rocket that produced it.

I should stipulate that in reality there are not one, but two calorimeters: first, the electromagnetic calorimeter, which essentially identifies electrons and photons, and behind it the hadronic calorimeter, which detects particles subjected to the strong interaction, such as protons and neutrons. (Some would blame me for not having mentioned it!)

But that's not all. You see, among the Little Thumbs, there are some who are really strong, and who resist Ogres (calorimeters): they are the muons, who interact very little with matter. So little that, in desperation, we no longer try to stop them with the Ogres, but more "simply" to locate them in the muon detector, which is on the periphery of ATLAS, and which is dedicated to them.

Thanks to this whole system, physicists are able to recognize the particles that pass through the detector, and to group those that come from the same collision by combining their energies and trajectories. Each collision can then be reconstructed by determining the particles that each one produced, as well as their trajectories and characteristics. So far so good!

But not everything is perfect in the best possible world of particles. Despite all the efforts made by the Ogres, some Little Thumbs completely escape detection: they are the neutrinos. In fact, they are Little Ghost Thumbs: they are both very light and invisible because they hardly interact with matter. Their existence was proved experimentally in specific detectors dedicated to this research. However, even if they are invisible in the LHC detectors, their presence can be guessed at because they carry with them some of the energy released by the collision, and this missing energy can be inferred.

To conclude concerning the ATLAS detector, we might say that it is a kind of gigantic camera, with a total of about a hundred million reading channels. Do you find this a small number, compared to the ten million pixels of your own camera? Certainly, but there are two major differences: the ATLAS reading channels are not all the same, and, above all, the trigger is 40 million times per second! No traditional camera is able to work so fast...

I think it's time to relax a little. Let's continue our visit to building 40 — it's worth it.

Building 40

We are in the circular hall: I like it because it is spacious and bright, with the cafeteria in the middle, of course.

The stairwells are arranged in a square, and on the wall of one of them you can see a huge image of the CMS detector — in real size! — composed of hundreds of "small" images put next to each other, like a gigantic puzzle that goes from the ground floor to the fourth floor.

The cafeteria in building 40

The large image of CMS

But, as luck would have it, this piece of art ended up on the wrong side of the circle. In fact, the building is divided into two parts: in the half-circle where the image is located are the offices of people working on CMS, while in the other half are the offices of people working on ATLAS. Thus, CMS people do not see the image of their detector, since it is on their side, and ATLAS people permanently see the image of the competing detector...

Discussions in the cafeteria

To be honest, I have never fully understood the nature of the relationship between ATLAS and CMS: rivalry, cooperation, emulation? It's hard to judge, it must be a quantum superposition of the three. I have an idea! We could ask the people we are going to meet today what they think. It will be fun, and we'll see what comes of it...

In any case, the ATLAS–CMS separation is given concrete form on the ground floor, where we are now, since there are tables and chairs on the CMS side, other tables and chairs on the ATLAS side, and in the middle the cafeteria. It's quite expensive, but once you've seen the prices of the main self-service cafeteria, you may come back here as fast as your legs will carry you. We will have the opportunity to have a cup of coffee here this afternoon.

There are four stairs, and for once their names are not very original: A, B, C, D. We will take stairway D, just next to the model of ATLAS. Before leaving the hall, do not fail to admire the circular glass ceiling above us.

The roof of building 40

Now, let's go to the first floor. The semi-open circular corridors that allow you to see the hall are very pleasant. I will show you my former office, where I worked for five months as an intern. Do you see the door ajar, with a strange light shining from it?

The door of my former office

It's moving to come back there. Wait, I think I hear a voice... Yes, it's Dimitris' voice, it sounds like he's on Skype. Let's listen, without being noticed...

"So basically you have a calibration for... partons? And another calibration for particles? Hmm. And when you look at figures 2a and 2b — 2.2b, sorry — we see that we have a different sign for Pythia vs Herwig... So for the particles we start with +10%, and it goes up to 1... while for the partons, it starts at minus 2%, and it goes up to 1. [...] So when you do a calibration for partons, you are more sensitive to *parton showering hadronization*. It is the *matrix element* that matters. Hmmm... Is what I'm saying true or not?"

Well, I think we'll have a hard time keeping up with what Dimitris says... In any case, nothing has changed in the office; the large planisphere and the large poster of the Chagall exhibition are still there. It's still office number 12, and next door... Right next door... Well, yes, it's still number 16. Even here the numbers have trouble following each other.

Ah, do you see all these posters hanging on the walls, called "Higgs Hunting"?

Posters for the *Higgs Hunting* Conference

This is an annual three-day conference on the Higgs boson. I attended it in Paris just before arriving at CERN, and I must admit that I didn't understand everything. I mean, I understood all the sentences like *Hello, everybody, I'm so happy to be there* or *Thank you very much for the invitation*, and I applauded warmly at the end of each talk. The only remark I would have been able to make is: "I don't know what this plot represents, but it is very pretty; the colours are really well chosen — well done." In fact, I really like the design of these posters, which have a masterpiece in the background. And on the 2016 poster, Gustave Courbet's *Le Déspéré* (*The Desperate Man*) seemed to me well adapted to my situation.

Let's continue a little. In this part of the building, several offices belong to the French laboratories, it's a bit like the national headquarters here! Well, I think I hear two French voices in this office. Let's approach without making any noise, and listen...

Needless to say. Frankly, the answer was in the question, wasn't it?

Instead of trying to understand what is being said in these offices, let us walk along the corridor. Since it forms a circle around the hall, I suggest taking a complete tour for pleasure. We are still in the half circle dedicated to the ATLAS detector. Look to the right: on this side, there are open spaces with a view of the hall... and more particularly of the huge image of the CMS detector.

Let's come back and take a peek through the windows of the offices on the left side. It's even ajar; we can enter discreetly... I would like to show you something quickly... Have you noticed?

Walls are no longer walls: they are whiteboards, with scribbles almost up to the ceiling. Pretty original decoration, isn't it? And, moreover, changing over time, along with ideas! You may also notice the presence of blue helmets on the shelf. Don't suppose that physicists in building 40 are UN soldiers in their spare time. It is simply always useful to have these helmets available, since they are mandatory in certain areas. You will have first-hand experience of this at the end of our visit, when we go down into the cavern of the ATLAS detector.

Let's go to the first floor. We arrive at stairwell B: that's it, here we are in the CMS camp. Here, let's take a little detour to this recess behind the stairwell, next to the elevator. Have

An office facing the CMS image

Inside an office

you seen the new generation of computers available at CERN?

Do not make fun of it, please. It seems that there is a certain taste here for the old-fashioned and its timeless charm. After this very useful detour, I suggest we return to the main corridor. I would like to show you a poster... We will discuss this in more detail this afternoon, but you already know that the Higgs boson was discovered at CERN. Well, the Higgs boson is one thing, but it's not all, there's also... THE PIGGS BOSON!

I must confess that I giggled the first time I saw that rather crazy poster. If you were to ask yourself by which treacherous stratagem the *Angry Birds* phenomenon could creep into the CERN compound, you now have the answer. Aren't these little Piggs Bosons cute, though?

We have almost done a complete tour and we are now back on the side of the French HQ. This is the perfect time to talk to you about the French contribution to the ATLAS detector. But I'm not going to do it myself: we're going to talk with Isabelle Wingerter, whom I see below and who has seen us. While she is coming up, I can tell you that she was ATLAS–IN2P3 coordinator — she will tell us what it

Old computers

is all about. In any case, she is very well placed to tell us what the French laboratories involved in the ATLAS experiment have achieved, as well as the prospects for improvements for the detector. Here she is.

ATLAS France

"Hello, Isabelle! The timing is perfect, we have just finished our brief visit of building 40.

- All right, hello! Shall we move into the office?

All right. So, first, could you tell us what ATLAS France is?

- ATLAS France is the grouping of all the physicists, engineers and technicians of the laboratories involved in the ATLAS experiment in France. I coordinate the laboratories that are part of the IN2P3, that is, from north to south: LPNHE in Paris-Jussieu, LAL in Orsay, LAPP in Annecy, LPC in Clermont-Ferrand, LPSC in Grenoble, and CPPM in Marseille. These six laboratories, therefore, constitute ATLAS-IN2P3, and include 150–200 people. If we add the CEA in Saclay, which is not part of the IN2P3, the seven laboratories constitute ATLAS France.

But, in fact, what is IN2P3?

- The IN2P3 is an institute whose acronym stands for Institut National de Physique Nucléaire et de Physique des Particules, that is to say National Institute for Nuclear Physics and Particle Physics; but instead of saying INPNPP, we express it as IN2P3."

Hard to guess!

"And what is the role of the ATLAS-IN2P3 coordinator, so your role now?

- Essentially it is to coordinate the activities of the various laboratories in order to improve the ATLAS detector. It is also to negotiate budgets with the IN2P3 management, discuss their distribution, and represent the IN2P3 in the ATLAS collaboration.

How did the ATLAS France laboratories contribute to the construction of the ATLAS detector?

- The majority of the French laboratories have joined forces to build almost half of the ATLAS electromagnetic calorimeter. The one in Clermont was involved in the construction of the hadronic calorimeter. Another laboratory that was particularly involved in the development of the pixel detector is the one in Marseille, the pioneer in this field. In addition to this, the CEA in Saclay participated in the construction of the muon chambers, and had the idea of the air toroid..."

This is what generates the magnetic field for the muon detector.

"... All this work was carried out between 1995 and 2005.

Can we say that the French are in a way the "champions" of the electromagnetic calorimeter?

- Ohhhh yes! I think we can say that! In particular, it was a physicist from the LAL in Orsay, called Daniel Fournier, who proposed the accordion shape for the calorimeter.

Why is this geometrical form so advantageous?

- The great virtue of the accordion calorimeter is that it is completely hermetic in the azimuthal plane..."

This is the plane perpendicular to the axis of the beam.

"... No particles can escape us because there is no blind spot. But of course, it was not easy to build.

How did you contribute to the electromagnetic calorimeter?

- I started in 1991 with tests in Prévessin on prototypes, which got bigger and bigger..."

We will soon go to Prévessin to see what a test on a prototype looks like.

"... In 2001, a full-scale calorimeter module was tested. And in 2004, we added a pixel detector in front of it and muon chambers behind it, to complete the overall structure of the detector: we really built a major portion of the ATLAS detector!

So, now that all these elements have been installed in the real ATLAS detector, let's talk about the next steps: what are the improvements planned for ATLAS on which the French laboratories are working?

- In fact, there are two periods of upgrades to come. Phase I of the upgrade will take place shortly, in 2019–2020. The teams involved in this phase are mainly Orsay, Marseille and Annecy,who will improve the calorimeter trigger system by better isolating the particles we want to select..."

The trigger system makes it possible to filter the signals in order to keep only those that are really interesting — I'll tell you more about it in a little while, don't worry!

"... Only the electronics are involved. We are at the end of the research and development phase, and we are starting to manufacture the first prototypes. This is for the first phase. For phase II of the upgrade, i.e., for the High Luminosity LHC planned for 2025–2026, and which will operate until 2037, the majority of ATLAS France laboratories are working on the construction of a new pixel detector, the ITK.

What will this new pixel detector do better than the current one?

- Well, first, the current detector will have to be changed because it will be too old; it will have received too much radiation related to the current data taking. Moreover, the new detector will be more efficient, of course! The big challenge is that with the High Luminosity LHC, we will have many more particles to detect: several thousand particles, 40 million times per second..."

Compared to 1000 now, in order of magnitude.

"... So we're going to have very small pixels, about 50 micrometers on each side, so that we don't detect two particles at the same time in the same pixel. Furthermore, for phase II, we will also have to change all the electronics of the calorimeter, because it will also have become too old.

And if we look even further ahead, in 2025, will we still have to think about other improvements?

- Yes, it will probably be necessary again in 2030 to change the pixel detector, which may deteriorate because of all the particles it will receive. The future pixel detector is designed so that the internal part can be removed and replaced.

I would like to ask two more personal questions... First, you are about to become Deputy Spokesperson in March, for a two-year term: it's an important position! Could you explain to us what it consists of?

- I will be one of the two deputies of the ATLAS Collaboration Spokesperson, who each have a very specific coordination task in the ATLAS experience. For my part, I will be responsible

for everything related to the operation of the detector and the preparation of the data for analysis. The other deputy is responsible for physics, triggering and outreach. ATLAS is a nice team, but not at all easy to manage: there are 182 institutes in 38 countries representing about 5000 people.

Last question, the most difficult... What do you think of CMS?

- Oh well... I think that's very good! Hey hey! They have a detector very similar to ours, with defects that compensate for its qualities, and which works very well. CMS collaboration is very rich on the human level...

Well, come on, there's a little rivalry between ATLAS and CMS, right?

- Yes, obviously, I want ATLAS to be ahead of CMS; if we can find something before them, that's good! But this rivalry is not negative at all: there is healthy emulation between us. And some teasing! We always have fun criticizing them a little.

Especially in the cafeteria, I guess! Thank you, Isabelle, for having told us about the contributions of the French laboratories; it somehow humanizes this technological monster, the ATLAS detector, which we will discover at the end of the day. See you soon!

- It was a pleasure. Goodbye!"

I think that this discussion with Isabelle gave you a good overview of the French know-how involved. What time is it? Oh, let's hurry, we're going to be late for the ATLAS Weekly! The ATLAS Weekly is the weekly meeting of about one hour that allows one to keep up to date on what is happening within the ATLAS collaboration, to take stock of the progress of each group. And when I say ATLAS collaboration, I mean ATLAS collaboration: even if it happens that we mix with CMS people in the cafeteria, no CMS member is invited to this meeting! You are therefore bound to secrecy: I am counting on you to remain discreet about what we are going to hear, in order to preserve the independence of the two experiments.

We need to go downstairs, follow me. Here we are at the amphitheatre level — there are four, as many as there are stairways, and their names are...? A, B, C, D, do I hear you suggesting? Almost! That's the idea, but to make the amphis a little less impersonal than the stairs, they were baptized: Anderson (the one who experimentally proved the existence of the positron, the antiparticle of the electron), Bohr, Curie and Dirac. I bet if there were an E, it would have been Einstein. (From that point of view, other physicists with an E, such as Edison, Ehrenfest or Everett, have been unlucky.)

As a general rule, the Anderson and Bohr amphitheaters are on the CMS side and therefore reserved for CMS meetings, whereas the C and D amphitheaters are on the ATLAS side and therefore reserved for ATLAS meetings. This time, our meeting takes place in amphitheatre D.

Fortunately, it has not yet started. Quick, take this place! Perfect. Yes, the seats at the far left or at the far right are very popular, because they allow you to escape stealthily... We're pretty comfortable here, aren't we? On the other hand, I am still mad at the person who designed the small writing tables, which are very small! I'm filing a complaint against X because there's no place for anything on them, and I had to go through acrobatic contortions during my internship to put the computer on my legs.

Traditionally, the first person to speak, for 5–10 minutes, is the ATLAS Spokesperson, or "Spoke" for close friends. Isabelle, whom we just met, will become assistant to the next Spokesperson, the German Karl Jakobs, who will take up his position in March. The current Spokesperson for ATLAS is the Englishman Dave Charlton. Then, for the other presentations, the speakers rotate from one week to the next. Here we go, I'll keep my voice down so I don't disturb anyone. The Spokesperson's presentation is about the news of the week.

"Well, let's start... Hello, everyone. So... Last weekend we started taking new measurements after the MD4..."

MD means "Machine Development." This is a period during which the LHC technicians, i.e., not of the detector but of the entire accelerator, study the beam to improve the performance of the machine.

"... So we are back to the usual measurement routine, and this will last

The ATLAS Weekly meeting

until the end of October, I will remind you of the schedule in a moment. This weekend we acquired data at low mu..."

So, since this weekend, the beam condition has become "normal" and conventional measurements have been resumed; "at low mu" means, roughly speaking, at low intensity, the intensity being proportional to the number of protons sent into the machine.

"... There are still about two weeks left of pp collisions..."

pp: did you guess it? It means proton–proton.

"Three weeks from Thursday there will be the MD5, then the following week the TS3, and then we will begin the run on the ions..."

So in three weeks there will be successively a new period of Machine Development, and a "Technical Stop." During such a technical stop, there is no beam at all; the LHC technicians "turn off the power" to solve problems that affect the entire accelerator. Of course, no more collisions, no more data! And by the end of the year, we will no longer collide protons, but heavy ions.

"... The ion data acquisition will start at the beginning of November, precisely on November 10, just after the four days of setup. And then it's the technical stop of the end of the year, until Christmas. That's the planning. Now some news from the physics coordinators: here is a summary table of the papers published this year for Moriond and EPS. I remind you that for Moriond, they were mainly concerned with research on BSM, and for EPS with precision measurements. We have a fantastic number of results! With a total of 186 publications..."

Moriond and EPS are two major annual international conferences where the particle physics community meets to present the latest advances. In particular, this is where ATLAS discovers the results of its CMS competitor, and vice versa! As for BSM, "Beyond Standard Model," it concerns the research for a new physics, which I will talk about soon.

"... This week is dedicated to the TRT, the pixel and the Tile. As you all know, next week is ATLAS week; there will be a large number of lectures: please attend as many of them as possible! I would also like to point out that there will be a poster session on Monday. Think about it — it is a good opportunity to present your work: work on the detector, improvements, triggering, computer calculation, software, outreach... We support all projects!"

Roughly speaking, this week is devoted to the pixel detector I told you about in front of the ATLAS model, as well as the calorimeter that identifies hadrons, so in particular protons and neutrons. It's different every week. And three times a year, including next week, there is the "ATLAS week," where everybody working on ATLAS meets to discuss each other's work.

"... Are there any questions, or comments?... No? All right. Thank you for your attention."

No, no! I sensed that you were going to applaud — that's not customary here, except for presentations by guests from outside the collaboration. The lady who is about to speak now will take charge of the second traditional talk after the News, namely the "Run Coordination Report." She will explain what happened in the different parts of the detector last week.

"Well, hello everyone. Last week was largely occupied by the MD..."

Machine Development, as before.

"Here you see a graph of the luminosity of ATLAS, during Machine Development from last Tuesday to Friday, during the increase in intensity on Saturday, and in stable beam since Sunday. In total, last week, the LHC delivered 960 bp^{-1}..."

This figure broadly corresponds to the amount of data that has been collected in the last seven days.

"... This gives a total of 32.9 fb^{-1} provided by the LHC and 30.4 fb^{-1} recorded by the detector since April, giving ATLAS an efficiency of 92.4%..."

This means that since the LHC was restarted in April, ATLAS has not recorded all the collisions: some protons are getting lost! In a perfect world, the detector should take into account all collisions, but this is not quite the case.

"... I will now review the different subsystems. An attempt was made to perform a high voltage scan of the IBL during the MD, without success..."

She's talking about the pixel detector. The IBL is the part of the pixel detector closest to the collision point, with 12 million pixels. In other words, this is where the Little Thumbs deposit their first stones.

"... For LAr, well, we are slowly returning to physics measurements, with the increase in luminosity. As for the Tile, we did a small operation during the MD to calibrate and test the new code for luminosity..."

LAr means "Liquid Argon," which is used to make measurements in the electromagnetic calorimeter. And Tile is related to a part of the hadronic calorimeter.

"... Concerning the muons: we've had a lot of bursts, 22 in 20 hours, we really need to look into this problem as a priority..."

She's talking about the muon detector, you remember, those particles that interact very little with matter. Apparently, they had faulty measurements.

"... So in summary, last week, we had 5 days of MD. Then the low mu run during the intensity increase period on Saturday turned out well, marking the end of the 2016 low mu programme. And since Saturday, we have been back to producing stable beam physics data. Thank you for your attention."

We're at a pause between the end of a presentation and questions. I think we've stayed long enough for you to get an idea of what ATLAS Weekly is. Let's go.

Take a deep breath, it's over! Even if it was not easy to understand, I wanted to take you to this meeting, which is a must-see moment. We have already talked about the detector; however, we have not yet mentioned an important aspect: the trigger system. On our way to Prévessin, I'll have time to tell you about it. We must already go to Prévessin again to attend a "test beam," i.e., the test of a new prototype that is part of the

ATLAS detector improvement programme. I also have a little surprise in store for you... We are leaving building 40 but rest assured, we will soon have the opportunity to come back to it. Where did I put the keys? Here they are, let's go!

Have you seen films like *Troy* or *The Lord of the Rings*, or series like *Rome* or *Game of Thrones*? There always comes a time when you first see two huge armies facing each other, and then their leaders in close-up with a piercing look full of hatred. After a moment of absolute silence, these leaders begin to scream and run like savages while brandishing their spears, followed by their respective armies: in general, a wide shot shows the two angry human masses rushing together to kill each other.

This is exactly the state of mind of protons when they arrive in the detector: accelerated to 99.999999991% of the speed of light, they are ready to fight. But — this is the notable difference from the armies in *Troy* or *Rome* — the protons are pathetic soldiers. In a subatomic army, that is to say a packet of about 100 billion protons, only about 20 of them collide head-on with a proton coming in the opposite direction; the others only cross paths with protons from the other side. When a high-energy collision happens, new small particles are created, then head off in all directions, and the signals they produce are collected everywhere in the detector. But the vast majority of these signals are unusable or uninteresting, so you need someone to put this myriad of signals in order.

This "someone" — you guessed it — is the trigger. Its goal is to sort the signals almost in real time in order to keep only the collisions (or events) that are exploitable and of interest to physicists. Considering that two packets of protons arrive inside the detector and collide every 25 nanoseconds, in 1 second there are — I'll let you do the calculation... There are... 40 million two-packet meetings, or "bunches." However, from what I was just telling you, a meeting of two packets produces about 20 collisions. A small multiplication allows us to estimate the number of collisions per second at 1 billion. Indeed, 40 million times 20 equals 1 billion. Don't tell me I'm bad at math — it's called an order of magnitude; for a physicist, this multiplication is perfectly correct.

Among the 40 million packet encounters, the trigger finally keeps only about… 400 of them. Signal filtering must therefore be done quickly! To make such a selection, you have to be in rather good shape.

After having carried out this first drastic sorting, according to criteria defined by the physicists for the needs of the experiment, the trigger will seek ROIs, i.e., "Regions of Interest." It tries to recognize in the measured energy deposits *candidates* for known particles: electrons, photons, muons, etc. Then it will take a few milliseconds to refine the analysis and determine if the candidate signals for known particles really correspond to known particles. Finally, it will send the 400 or so selected events to the Computing Center, the CC. This is not to be confused with the CCC, the CERN Control Centre, which we already visited. You could be blamed for a mistake on the number of Cs — be careful. But the Computing Centre is another subject we will talk about this afternoon.

Let's continue this part of our visit dedicated to ATLAS. Now that we have an overview of it, I would like to show you in the field how we test new technologies to improve the ATLAS detector. That's why we will visit a large hangar in Prévessin, where laboratories can test the performance of new components by subjecting them to a beam of electrons or other friendly particles. These tests are very important to guarantee the quality of the components that are intended to be installed in the real detector.

Let me first remind you of the programme for ATLAS, which Isabelle Wingerter detailed earlier: we are currently in Run 2, i.e., in the second phase of data acquisition for the LHC. After a technical break that will last six months, we will move on to Run 3, which will not be fundamentally different from Run 2, except that the luminosity will be slightly increased. Then there will be a new technical stop, but this time to drastically increase the luminosity of the LHC. This means that there will be many more collisions: we said that today, when two packets of protons meet in the detector, they produce on average about 20 collisions. Beginning in 2030, during the High Luminosity phase (High-Lumi), a meeting of two packets of protons will produce something like 200 collisions. The more collisions, the more Higgs bosons are produced: good news! But it is a real challenge for everyone, because downstream we need efficient technologies capable of managing all these collisions at all levels: accelerators, detectors, trigger…

We are approaching the entrance to the Prévessin site, please take out your visitor's card and show it to the guard. Laurent Serin, whom we will soon meet, will introduce us to technology precisely designed for the High Luminosity phase. I think he will also tell us about the upcoming deadlines. I can't wait to find out what he is cooking up in his hangar.

Test beam in Prévessin

I see Laurent waiting for us at the entrance of the large hangar. Laurent formerly had a special position as Scientific Assistant Director in charge of particle physics at IN2P3. But after a few years of administrative and budgetary tasks, he should not be dissatisfied to be back on the ground! Let's go join him.

"Hello, Laurent, and thank you for welcoming us to the test beam hangar!

- Hello! Wait, you'll thank me at the end maybe, not now. As we anticipated, there is no beam today, so we can walk in the hangar safely. Follow me, please."

This hangar is really huge! Look at all this concrete to protect the staff from parasitic radiation... That's impressive.

The test beam hangar

"Here we are in the hangar dedicated to test beams, that is, to the tests on prototypes subjected to a beam of protons or other particles. These prototypes contain new technologies that are intended to be used in the ATLAS detector, if the tests prove conclusive. Let's go up to the gangway to get an overview of the hangar.

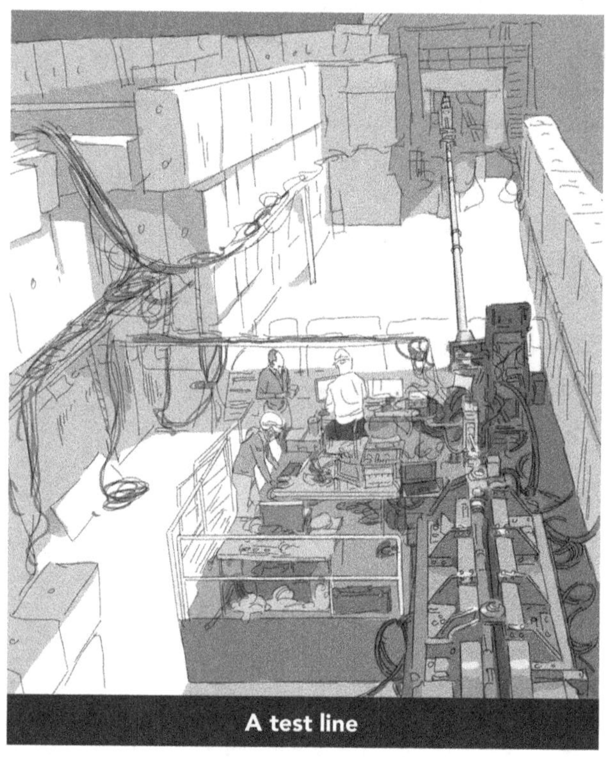

There are four independent parallel beams that cross the hangar from one end to the other in grey tubes, like the one you see below at the bottom. On each line, portions of the tube are missing in certain places: this is where the physicists place the samples they want to submit to their particle beam in order to test them. So there are areas in this hangar where the beams circulate in the open air; but don't worry — when the tests are in progress, there is no one nearby. Over there, at the very bottom, each line is supplied by the SPS..."

The SPS is the Super Proton Synchroton — remember? It is the insertion ramp that accelerates the protons before sending them into the LHC. If the majority of the protons accelerated by the SPS then move into the LHC, some of them are used for the tests performed here.

"... The experiments on this side of the hangar, at the starting points of the different lines, are experiments that do not deteriorate the beams; the particles lose a little energy as they pass, but it is insignificant. These are tracking experiments..."

So these are experiments that correspond to the pixel detector.

"... while on the other side of the gangway there are calorimetry tests that absorb the particles of the beam, except the muons. That's why the muon experiments are at the very bottom on the other side of the hangar."

The chain of tests carried out here therefore follows the principle of a detector: first the pixel detector, then the calorimeters, then the muon detector. Several people use the same beam, so you have to share it intelligently!

"... Our experiment is right underneath.

Can we go down to see more closely?

- Yes, come with me. When we are in the middle of the High Luminosity phase of the LHC around 2030, when two packets of protons meet in the detector, they will produce about 200 collisions. A collision releases new particles whose trajectories are measured, which makes it possible to extrapolate to the location where it took place. The problem is that because of the large number of collisions that will occur in 2025 in the same encounter of two packets, some of them may happen in the same place but at different times..."

That is true: for example, at a given point, there can be a collision of two protons at time t, and then another collision with two other protons at time t + a very small amount.

"... in other words, it will not be enough only to find where the particles have collided, but also at what precise moment, in order not to mix particles from different collisions over time. Especially at the ends of the detector, in the High Luminosity phase, it will be necessary to be able to distinguish particles that come from the same place but not from the same collision. That is why we are testing new silicon detector technology with better time resolution, to know precisely when the particle passes through the detector, and to be able accurately to determine the location and time of the collision from which it originated."

Interesting! But don't you think we should ask for clarification on this resolution in time?

"This resolution in time you're talking about... What does that mean exactly?

- Well, right now, in the silicon pixel detector just next to the proton beam, we are only able to identify the passage of a particle at a specific location. But the trace we observe does not allow us to know exactly when the particle passed through this place. It's not very important in the middle of the detector, but at the extremities it's more frustrating, because you can't know from which collision the particle comes from.

In the black box you see there, which is placed along the axis of the beam, we put the prototypes we want to test. Ideally, we would like to have a time resolution of about 30 or 50 picoseconds, that is, 30 or 50 millionths of a millionth of a second: if we can determine when the particle passes with an accuracy of 50 picoseconds, we can deduce from which collision it comes."

In short: at the ends of the detector, it is not enough to know *where* Little Thumb left his stone; it is also necessary to know precisely *when*, to be able to find *where* and *when* the original collision took place.

"Last week we tested 1 mm² silicon prototypes, and yesterday we changed to 3 mm² pieces. We are waiting for the beam to come back on Thursday to start new data acquisition, which is done — move forward a little — by this oscilloscope, which is itself controlled remotely from the control room above.

3 mm²? It's very big for a pixel, isn't it? In the pixel detector, it is normally much smaller, something like 100 microns...

- Yes, but this is in the centre of the detector. At the extremities, we have many fewer particles passing through, so we don't need to have such high precision.

All right! But let's get back to more down-to-earth things: what's the schedule?

- The schedule? It's simple: by the end of the year we have to prepare an IDR, an Internal Design Review, to start presenting our work to the ATLAS collaboration. Then in June, we will do a Comprehensive Review, and this will be the moment of truth: either the collaboration accepts the implementation of our technology, and we are happy with an approved project, or... we have to do something else! But if we remain optimistic and everything goes well in June, by the end of 2018 we will have to write a TDR, a Technical Design Report, to prove that we will be able to build our sub-detector with the required performance and on time for the High Lumi.

But I imagine that all these steps will be completed without any problem...

- Not at all — you're kidding! Besides, I didn't tell you about all the trouble... First, the realization of these doped silicon plans for our needs is really complex, the process involves 80 steps. It is handled by a company based in Barcelona. Then, we have to prove with our tests that our silicon plans are resistant to neutron radiation, which is created everywhere, and which is disastrous because it can induce a loss of doping in silicon, and therefore a loss of signal. They must also resist ionizing particles that deposit energy in silicon and damage electronics. A "0" can become a "1", and it's a catastrophe... Showing that our technology is robust is one of the objectives of our tests. A whole programme in itself! Shall we go up?

Yes. One last question before we leave... According to you, is there a rivalry between ATLAS and CMS?

- No... I don't like the word rivalry. We are competing for results, that's true, but there is no negative aspect, which is what the word rivalry implies. Of course, if we can find a good result before them, we will do it! And it's reciprocal.

So it's a friendly competition. Thank you very much, Laurent, for presenting your work in this huge hangar. See you soon!"

Following our meeting with Isabelle Wingerter, who told us about the involvement of French laboratories in the ATLAS detector, I think it was interesting to meet Laurent to hear about test beams, in order to discover a little bit of the reality on the ground. It also gave you an example of a challenge to be met in the perspective of the High Luminosity LHC.

We will now take a break with a little surprise that I promised you earlier... In fact, we're not very far from it — follow me. My surprise is in this building across the street. We're now going into space... Without further introduction, this is the building that houses the control room of the AMS, the Alpha Magnetic Spectrometer!

The ATLAS control room

The ATLAS mural fresco

The auditorium seen from above

Building 40

Cafeteria in building 40

The CERN Control Centre logo

Corridor in building 40

Glass corridor near the Computing Centre

The shell of a dipole magnet, near the restaurant

The exhibition "Universe of Particles"

Niels Bohr Road leading to building 40

The Globe of Science and Innovation

The landscape seen from Niels Bohr Road

LINAC 2

Part of the ATLAS detector

The test beam hangar

The Web Corridor

The AMS, Alpha Magnetic Spectrometer

I must make amends, because since the beginning of the visit, I have been full of praise for the LHC, which I claim is the largest particle accelerator in the world; in fact, that is not true. There is a much bigger and much more powerful one... The universe itself! Indeed, the universe is traversed by a flow of very energetic particles, whose origin remains uncertain, and which constitute what is called cosmic radiation. It is composed essentially of protons, but also of antiprotons, electrons and positrons, helium and lithium atoms...

The building dedicated to the AMS

Why is it important to study these cosmic rays? Because they are far from having disclosed to us all their secrets, and because they may carry answers to fundamental questions such as: why did matter prevail over antimatter in the early universe, or what is dark matter? The problem is that it is difficult to study cosmic radiation from the Earth's surface, because the atmosphere is an effective shield. So, what could be better than sending a particle detector... into space!

This is precisely the project that was launched in 1995 by NASA in collaboration with CERN, NASA having recognized the scientific interest of the project. Why NASA and CERN together? For a very simple reason: because NASA is an expert at sending devices into space, and CERN at manufacturing particle detectors! After years of design and testing, the AMS 02 detector (Alpha Magnetic Spectrometer 02) successfully docked in May 2011 at the International Space Station, at an altitude of approximately 400 km.

I decided to tell you about this experiment, which is absolutely unique in particle physics, because it takes place in space and thus blends with astrophysics. But why have I decided to tell you about it now? Because it is very interesting to see that AMS and ATLAS, although one is scanning cosmic rays in space and the other is studying particle collisions 100 m underground, are looking for answers to the same questions and have comparable detection methods. So, following the part of our visit devoted to the ATLAS detector, I didn't hesitate to bring you here!

Let's get closer to the model of the AMS on our left. It is on a 1/3 scale: the real AMS is 64 cubic metres in volume, weighing 7,500 kg.

What do ATLAS and AMS have in common? At the centre of ATLAS is a pixel detector, which identifies particles that pass through it, and whose trajectory is curved by the presence of a strong magnetic field. You can see on the model of the AMS that it also has a pixel detector in its centre, consisting of 9 layers of silicon; in other words, the Little Thumbs from space deposit at most 9 small stones in the AMS pixel detector. It is surrounded by a magnet that creates a magnetic field inside, like the ATLAS solenoid magnet that I have already briefly mentioned.

Then, in ATLAS, there is an electromagnetic calorimeter that allows us to measure the energy of photons and electrons in particular. The same is true for the AMS, which has an ECAL electromagnetic calorimeter at the very bottom of the

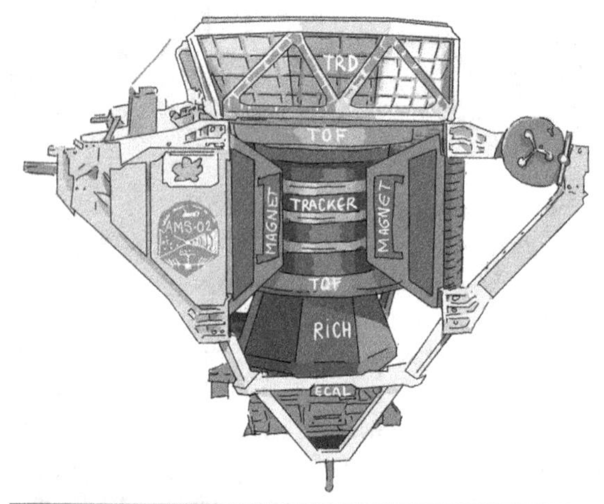

Model of the AMS

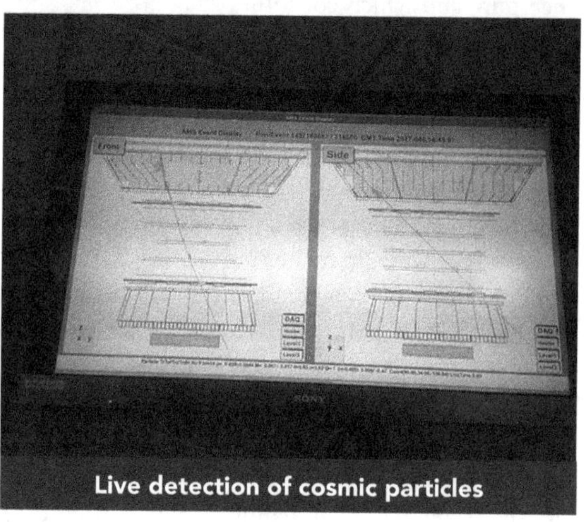

Live detection of cosmic particles

model, and which allows the energy of electrons, positrons and photons to be measured.

Another common point: like ATLAS, AMS has a trigger that chooses interesting events, on average 500 per second. For example, particles that do not pass in the axis of the detector, i.e., those that do not go "up and down" with respect to the model, are not selected.

But given the very high energies of cosmic radiation particles, AMS has other specific sub-detectors. For example, you can see on each side of the pixel detector the TOF (Time-Of-Flight), which notifies the arrival of cosmic rays to the other sub-detectors; at the top the TRD (Transition Radiation Detector), which recognizes electrons and positrons; and at the bottom the RICH (Rich-Imaging Cherenkov Detector), which identifies very fast particles and measures their speeds.

Now, turn around and look at the screen in the corner of the room... We can follow the detection of cosmic particles live!

Between its launch in 2011 and 2016, AMS recorded approximately 85 billion events. In particular, data analyses have shown an intriguing excess of positrons in cosmic radiation, compared to theoretical predictions: is the explanation related to the possible existence of dark matter, the unknown substance that could make up a quarter of our universe? Another perspective to consider: are there antimatter atoms in cosmic radiation, such as antihelium? No one is yet able to answer these questions, but one thing is certain: the AMS experiment, which may lift the veil on the great mysteries of the universe, has beautiful years ahead of it!

Before we leave, look behind that window. This is one of the two control rooms of the AMS, the other being based at NASA. The data from the detector is first collected by NASA, which then transfers it here to CERN for analysis.

Did you notice that large armchair at the end of the table in the centre of the room? This belongs to Samuel Ting, Nobel Prize winner in Physics and Scientific Director of the AMS. It seems that it is forbidden, even in his absence, to sit on it...

After this escape into space, I think it's time to come back to Earth. Let's get back to the car: we'll go back to the Meyrin site. You can say goodbye to Prévessin — we probably won't come back here.

We have arrived by slow steps at a crucial moment of the visit: lunch! Let's head back to building 40: CERN's main restaurant is right next door. I'm going to park a little further

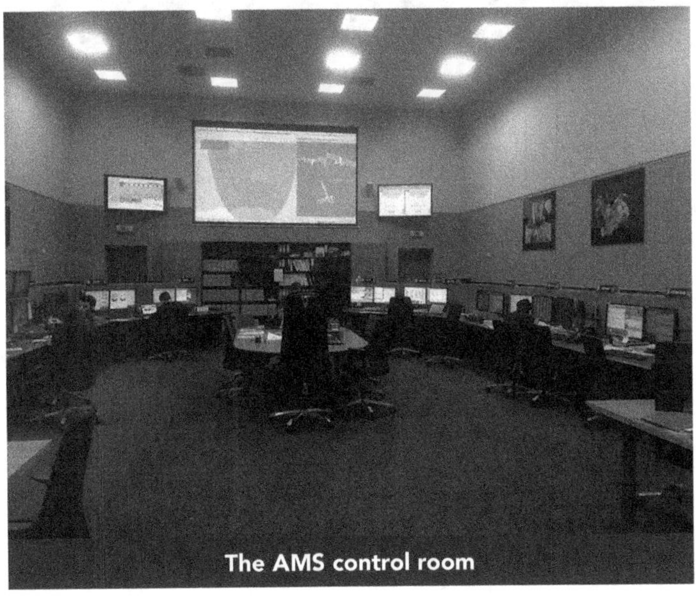

away than I did before. The architecture of building 40 is also impressive from this angle, isn't it? From here, this building has the shape of a magnet, with the inner coil and the breech!

Lunch will be an opportunity to come back to the protons we left earlier in the ATLAS detector. As you know, their collisions produce new particles. Among these appears sometimes... the Higgs boson, which has recently entered the bestiary of known particles. I will talk to you about it during the meal, after which we will resume our journey by discovering the buildings near the restaurant, which, as you will see, are worth a visit.

The self-service meal

Let's cross the large lawn and head for the restaurant on the other side. Don't you find this expanse of greenery pleasant? It's fresh, it's pastoral, we breathe! I have only one regret: a swimming pool is missing... We pass by the relic of a blue cylindrical magnet a few meters long with "Accelerating science" written on it; its twin exists right next to the Globe. It is a dipole magnet, the one whose specialty is to guide protons. I can see that you're dreaming of taking one with you as a souvenir, but you might as well know that each costs about 500,000€. Besides, incidentally, it's 15 meters long and it weighs 35 tons...

The shell of a dipole magnet, near the restaurant

This blue colour is emblematic of CERN; you will find it in various places: on the CERN logo, on the banner of the ATLAS website... To shine in society, it's a very good thing to know; if you're looking for an original blue, instead of referring to Yves Klein's blue, you can mention CERN blue. To me, it's not lacking in style.

The tradition is to take a picture in front of the magnet. Stand there, don't move. Smile, we're at CERN! I'll send you that one. This photo is inevitable — you couldn't leave CERN without it. That's one good thing done.

We cross the terrace. We are lucky; the weather is fine and especially not too cold. We will soon find a seat there. It's a really nice place, with all these people sitting at tables talking about physics — but not only! — in all languages. It is always crowded at this time of the day.

Behind the bay window we're walking in front of, there is a large annex of the restaurant with tables arranged as a "sushi bar." Let's go

The terrace of the restaurant

in through that door on the right. Well, they've put a ping-pong table in the middle of the restaurant — why not?

Let's have lunch now. We are now at the heart of restaurant 1; there are two of them like this at CERN. Here we sometimes meet the Director General, Fabiola Gianotti, sometimes the spokespersons for the experiments or their assistants, and sometimes Nobel Prize winners... Look at the ceiling, too: it resembles a myriad of small stars, which steadily change colour. Being a particle physicist doesn't mean that you can't have your head in the stars!

You should be warned right away: here, it is very difficult not to give in to temptation, and the wallet suffers very quickly. You can easily understand that by looking at this dessert display: it's like being in a pastry shop.

Fruit salads, apple-crumble compote, mango or red fruit cottage cheese, chocolate mousse with sugar tiles, apple pies, Paris-Brest cakes, praline cakes, the speciality of the day: refraining from taking something is really complicated. Really complicated... Well... Here, I'll... I'll have a Paris-Brest cake. It's really not a good idea, considering the culinary surprise I have in store for you. I see you staring at the traditional red wine beef blanquette and the salmon tartare with lime, but these hesitations are useless, because you will not escape my surprise... Let's go to the bottom, where

Ping-pong players in the restaurant

Restaurant 1

The pastries

pizzas, pasta, burgers and grilled meats are available. You won't leave CERN without a taste of the Higgs not-to-be-missed pizza! You can choose between the Higgs boson Pizza with ham and cheese, like this one, or vegetarian.

Can you tell me what the ham and cheese represent? No? Well, don't you recognize a proton–proton collision? The two asparagus pieces facing each other in the middle are the two protons that get closer to each other, and BAM! Their collision gives birth to the chorizo at the middle of pizza (the Higgs boson), as well as the olives (neutral particles) and the curved ham slices (charged particles, which, under the influence of a magnetic field, follow a curved trajectory in the detector). And the chorizo itself disintegrates in this case into salami slices (i.e., photons). When you eat the mozzarella, you are eating the pixel detector and the calorimeter, while to savour the surrounding dough is to savour the muon detector.

The Higgs pizza

Now that you know everything there is to know about ham and cheese, you will have no trouble understanding the vegetarian recipe; it's not that different. Two asparagus-protons form, after their collision, a Higgs boson cherry-tomato, which disintegrates into four long slices of peppers (four high-energy muons), as well as smaller peppers (other particles produced at the same time as the Higgs boson). You will notice that the long slices of peppers (muons) are found in artichokes (the muon spectrometer). To be perfectly honest, the scientific significance of the Higgs pizza is rather dubious... But it's a delicious pizza — that's the essential thing.

Let's go to the till. You have only euros, I suppose? As you know, we are in Switzerland, so we should pay in Swiss francs. Don't worry, they accept euros, but give you change in Swiss francs. You may even notice that the currency in which you pay is not important. We have to pay a total of 25.50 francs: whether you give 30 euros or 30 Swiss francs, you will be given 4 francs 50 back. It is rather surprising to note such a lack of rigour here in these units, which are the cornerstone of physics and science in general! To be consistent with the spirit of the place, I interpret this as a quantum payment: the currency of the payment remains in a superposed euro–Swiss franc state until one has paid, and indeterminacy disappears when one hands over the banknote. No, don't pay, this is your first visit: you are my guest!

We should have lunch outside to enjoy the weather. On the way, take note of the "particle collision" tables!

When the weather is fine, it is a pleasure to have lunch in this restaurant. The view is beautiful; we will go up a bit higher later to admire it even more. In the meantime, *bon appétit*!

The "particle collision" tables

The Higgs pizza is a perfect opportunity to talk to you about the Higgs boson, which has been CERN's pride since its discovery in 2012. No, I swear, it really wasn't planned! Do not see it as a premeditated act, but rather as a happy coincidence.

Before mentioning the Higgs boson, as an introduction, let's first take a quick look at the other elementary particles already known. Since the 1960s, physicists have developed the famous "Standard Model," the reference theory in particle physics, which classifies particles into different categories.

First of all, you have the quarks, at the top left: they are the particles subject to strong interaction, the fundamental force of nature that ensures the cohesion of atomic nuclei. Two types of quarks, called "up" and "down," are the components of protons and neutrons that assemble to form atomic nuclei. The other quarks are much less familiar to us because they are much heavier and very unstable. Nevertheless, in addition to the traditional protons and neutrons, which are made up of three quarks, there are many possible combinations of two, three, even four or five quarks! In 2015, for example, the LHCb experiment signaled the fleeting but real existence of structures with five quarks, called pentaquarks.

On the right, you have the leptons: unlike quarks, leptons are not subject to strong interaction. There you find the electron, which really looks happy, the muon, which is still as shy as ever, and the tau, with its sly little smile. The muon and the tau are cousins of the electron, with the same negative electrical charge, but about 200 and 3000 times heavier, respectively. These three leptons are each associated with a neutrino, just next to them. Look how suspicious these neutrinos appear behind their black blindfolds... That's normal: these are the ones I called earlier the Little Ghost Thumbs, which the sensors can't identify. It has recently become known that neutrinos have masses that are not zero, but so small that they cannot yet be measured.

Then, at the bottom left, you have the messenger particles which carry the forces. Indeed, each fundamental force of nature is associated with a particle: the photon with electromagnetism, which governs the behaviour of charged particles; the gluon with strong interaction, which allows protons and neutrons to remain linked within the atomic nucleus; and the Z and W bosons with weak interaction, which is particularly responsible for a certain type of radioactivity. What about gravitation, you might say, and its hypothetical associated particle, the graviton? The gravitational force is much less intense than the others, and its effects are noticeable only when very large masses are involved. Its theoretical form is such that it's very difficult to fit it into the Standard Model. This is what gives rise to all the research on the theory of everything: physicists seek to bring together the three interactions of the Standard Model with gravitation to make a unique and complete description of nature. But that is another story...

ELEMENTARY PARTICLES

QUARKS

Up

Down

Charm

Strange

Bottom

Top
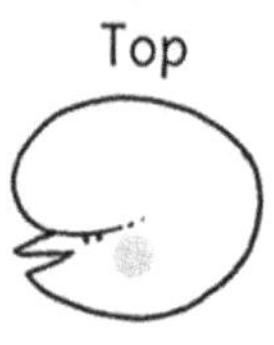

LEPTONS

Electron

Electron Neutrino

Muon

Muon Neutrino

Tau

Tau Neutrino

FORCE CARRIERS

Photon

Gluon

Z Boson

W Boson

HIGGS BOSON

With the majority of the elementary particles I have presented to you, you can associate an antiparticle, which doubles the number of particles. Finally, the Higgs boson recently entered the Standard Model: it is the one that gives mass to its congeners.

But let's go back to the time when the Standard Model does not yet contain the Higgs boson. This Standard Model, despite its efficiency, seems to have a flaw; or, at least, a quirk. It predicts that all elementary particles should have zero mass, which seems absurd. How can we react when a clearly very robust theory makes such a surprising and apparently meaningless prediction? There are actually only two possible reactions.

The first is to say that the theory is wrong and that it must be changed. This is called a legislative solution: you think that the laws set out by the theory are wrong because they do not seem to describe reality well, and that they must be amended accordingly. Here, this would mean modifying the Standard Model to predict non-zero masses for particles that are actually massive. It's the least we can ask for, isn't it?

But a handful of physicists, certainly more reckless, chose a second option. They decided to take the theory over the facts and experiment seriously, affirming that all particles do indeed have zero mass, and that we have always been wrong about the concept of mass.

Here is what they postulated in the early 1960s: mass is not, as one might think, an intrinsic (or fundamental) property of particles, but a secondary property. Thus, according to them, particles do not have mass according to the Standard Model, and they *acquire* it afterwards. Yes, but in this case, what gives mass to the particles? It is a field associated with a specific particle, a boson, baptized *a posteriori*... Higgs boson.

To solve the anomaly of zero mass particles in the Standard Model, Peter Higgs, together with Robert Brout, and François Englert independently, proposed not to modify the laws of the Standard Model, but to add a new ingredient: this is called an ontological solution, that is, one related to existence.

The principle is that elementary particles acquire mass by interacting with the Higgs boson through the Higgs mechanism. Here's an analogy: you know *Singing in the Rain*, don't you? In the film, Gene Kelly is thoroughly happy because he is in love. He is alone in the street because no one else is outside in this pouring rain, and he feels free; he dances, he plays with his umbrella, he jumps on a street lamp, in short: he feels light. He feels *light* because he does not interact with anyone; he is alone with his happiness, he twirls around.

Now, have you seen, for example... *Quantum of Solace*? In one scene in the movie, James Bond is led to pursue the villain in the streets of Siena during the Palio. The Palio is a horse race held in the main square of Siena, where ten horses, each representing one of the districts of the city, compete against each other; a huge and very dense crowd is gathered to attend the event. The villain tries to blend in with this crowd, and James, who emerges from a sewer mouth in the middle of the square, runs after him. But he is constantly stumbling against people, which slows him down, which makes him *heavier*. The same is true for the mass of the particles: the more they interact with the surrounding Higgs field, the heavier they are.

If you are a mountain dweller, you can see things differently. The quantum field theory, which effects the marriage between quantum mechanics and Einstein's special relativity, predicts that each particle is associated with a field; therefore, without much surprise, the Higgs boson is associated with a field, the Higgs field. Now imagine that this Higgs field is the snow of a ski slope: the equivalent of Gene Kelly would be the skier who runs down the slope without being excessively slowed down by the snow, and who is therefore light, while the equivalent of James Bond would be the snowshoe hiker, who at every step sinks into the snow and moves forward with difficulty — he feels heavy (even if it's hard to imagine James Bond in difficulty on snowshoes — I acknowledge that). I think this example of the snow is the one most frequently given when someone tries to explain what the Higgs boson is. In summary, the more the particle (the hiker) interacts with the Higgs field (the snow), the more it moves forward with difficulty, so the heavier it is.

I hope you enjoyed this Higgs meal, with its boson and pizza! We will soon start visiting the buildings near the restaurant. But first, let's stop for a moment in front of this rather surprising place. Water fountains, jugs, glasses and plastic cups are not exceptional; but the screen above, which indicates in real time the state of the beam that circulates 100 m under our feet, is much less typical!

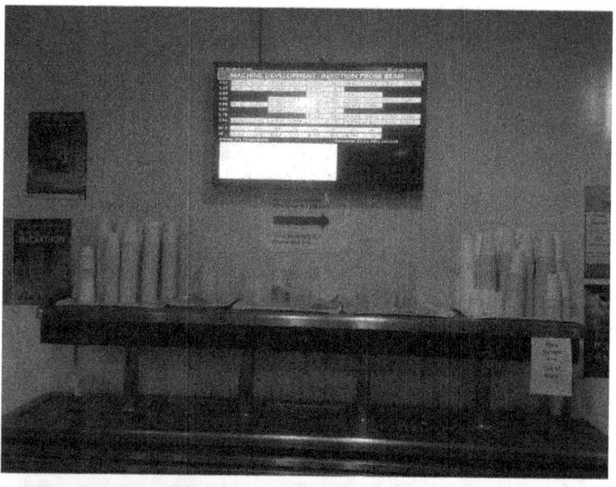

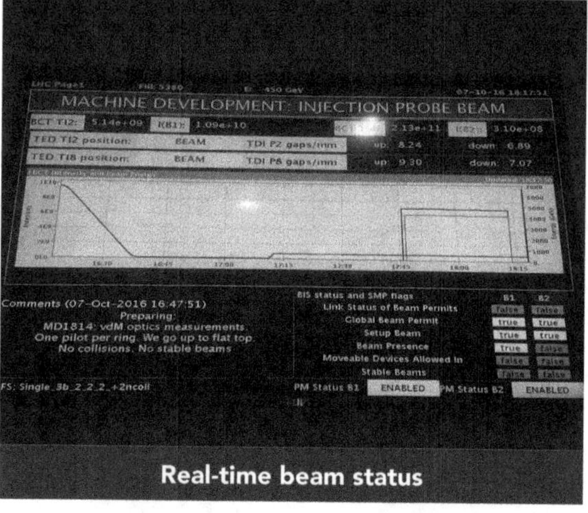

Real-time beam status

The curve in the middle represents the intensity and energy of the beam over the last two hours — more precisely, of the two beams in opposite directions. For the rest, you have a panel of other characteristics, with two columns, B1 and B2 at the bottom right, for both beams; B stands for Beam. There are comments at the bottom left of the screen — at the moment: "No collisions. No stable beam." This indicates that we are in a period of Machine Development, as discussed during the ATLAS Weekly meeting.

You will have noticed over the whole course of the visit that there are such screens everywhere. One thing is certain: CERN physicists live by the rhythm of the beams that circulate in the various accelerators. So here there is only one state of health that really concerns people, and that is the health of the beams. Don't worry, generally, they are doing well!

The Auditorium

et's get out of the restaurant. We are now in the heart of building 500, a microcosm all by itself. Mind the step. Immediately to the left, it is the tobacconist-news agent. You can even play EuroMillions at CERN: it is not in all companies that you can play EuroMillions.

This hall is pleasant, spacious. Here is the Post Office, there a travel agency, and of course, the UBS bank. Let's sit down for a while in front of this beautiful UBS stand.

The hall with the UBS bank

When I arrived at CERN for my 5-month internship, the question arose as to whether or not it was advantageous to open an account at UBS. So I went down the stairs behind the desk, and that's when I experienced the Swiss banker. I was suspicious, I expected him to try to use all possible arguments to push me to open an account, while remaining very friendly. My internship referent thought the opposite: she thought that they would probably discourage me from opening an account for such a short time, that they were only interested in big fish, and that they would probably be unkind to me. In fact, I found myself in front of a shy man, very nice, with a kind of permanent neutral smile. But the most surprising thing is that he was totally indifferent to the possibility of opening an account for me. I asked him a number of questions to help him convince me that I needed to open an account at UBS. No point — he was even more undecided than I was. So much so that I ended up asking him this question, incredible in the context: "But, according to you... Is it a good idea to open an account for these 5 months?" In my whole life, I never thought I'd ask a banker such a question. It's like asking a grocery store: "Tell me, do you think it's a good idea to buy fruits and vegetables in your store?" But the worst was the answer: "It's up to you — as you wish." Among other things, a personal consideration finally convinced me that it was better to open an account: it was the idea of having both a UBS card and a CERN badge. You feel a little in control of the world...

Let's continue our visit. At the end of the corridor behind the stairs is the User Office. A User is a person who is not directly employed by CERN, but who works on the CERN site for a laboratory. In fact, most of the people you meet here are not employed by CERN, but are only Users. As Bernard said this morning, there are about 2500 CERN staff members — the vast majority of whom are engineers and technicians — and 13,000 Users, who come to "use" CERN facilities.

I suggest going to the first floor. The characteristic Y-shape of this staircase always reminds me of the Grand Staircase in *Titanic*, doesn't it you? You know, the one where Jack and Rose meet at the very end, in a dream. Of course, this one is less luxurious; nevertheless, you will notice small garland lamps under the railing, which make a pretty effect when it is dark outside and when all the lights are on. Apart from the aesthetics, do not expect to meet romantic couples or movie stars, but rather, in general, particle physicists. Don't be so disappointed, it's not so bad! In any case, I didn't imagine that for five months of my life, I would think almost every day about *Titanic* for at least five seconds. I don't regret it: five seconds of romance a day is like five fruits and vegetables a day: it's very good for your health.

CERN's "Titanic" Staircase

As we go upstairs, look at this strange work of art above our heads...

What could it represent? A Cy Twombly scribble? The 3D erasures of an unsatisfied theorist? Dark matter convolutions? A threatening cloud hovering over physics, heralding an unprecedented revolution? I'll let you judge.

Now that we are on the first floor, I have to show you something absolutely essential, in the corner behind the door... I hope it's open... Yes, we are lucky! Come on. You are entering a historic place!

The work of art above the stairs

The auditorium seen from below

This is CERN's main auditorium, where the discovery of the Higgs boson was officially announced on July 4, 2012, in the presence of Peter Higgs and François Englert. Moving, isn't it? For my part, I remember feeling some chills the first time I walked into this room, which I had often seen on the Internet or in scientific journals. I remember that I said to myself at that very moment: "This is it, I'm definitely at CERN." Not many people have the chance to walk among these bleachers! Let's enjoy the moment, and admire the auditorium from above.

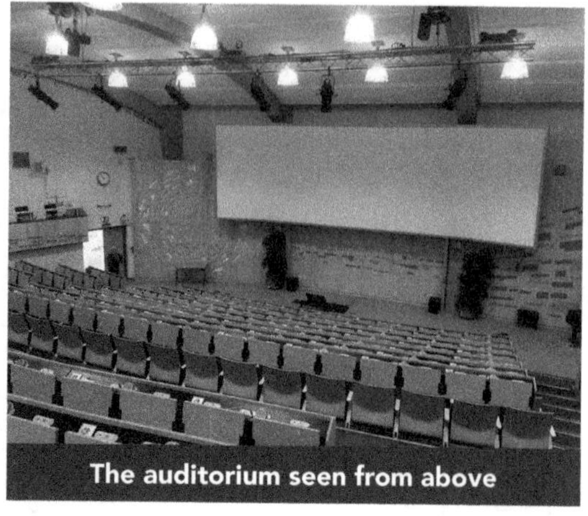
The auditorium seen from above

Have you noticed how high-tech it is? Almost every row has a microphone at every other seat, so you don't have to scream all over the amphitheatre if you have a question. There must be 400 places. It's big, but if you count only the coveted seats, the amphitheatre suddenly becomes very small. The places in question are those at the ends of the rows, preferably as high as possible. You can easily understand the reason, since I already pointed it out to you during the ATLAS Weekly meeting: these are the places that allow you to escape discreetly. The problem, when these seats are all taken, is that the people who are there block the way for all the others sitting in the row. We often find ourselves in an absurd situation, where people are desperately looking for a good place to sit and hesitate to sit down, although the amphitheatre is almost empty!

In your opinion, what do these motifs engraved in the wood between the door and the large white screen represent?

In fact, it is a detector. When you don't know that, it's not very easy to guess, I suppose!

Unfortunately, I was not in this auditorium on July 4, 2012, but I've been told about the atmosphere that prevailed there. The entire world knew that the announcement of the discovery of the Higgs boson would make this seminar historic, to the point that 200 courageous people slept in front of the auditorium to make sure they would get a seat. Admittedly, the auditorium may

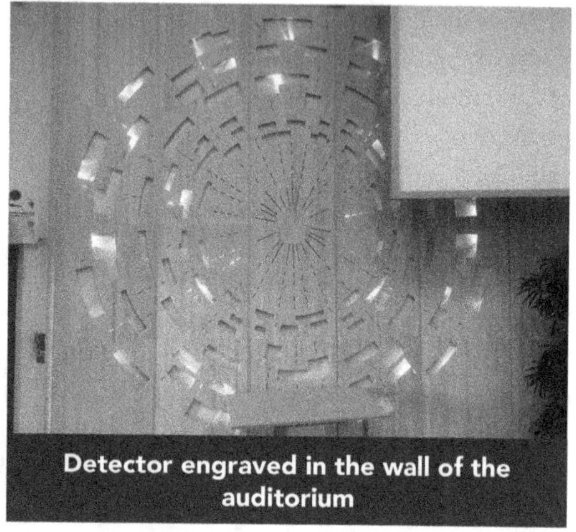
Detector engraved in the wall of the auditorium

seem large, when there is no one there; but at the time of this seminar, it was impossible to welcome thousands of people... Thus, apart from the VIPs who had their reserved seats, and the night warriors, the majority of people could not attend the conference in the auditorium and had to watch its broadcast.

I was also told that Peter Higgs, who was morbidly shy, did not really want to stay after the conference and tried to escape at lunch time, but he was no longer very young. He was harpooned by journalists and the entire scientific community, who rushed to him and prevented him from running away. On the other hand, it seems that François Englert is very talkative. I would have liked to be present at the meeting of the two men! It's the proof that we can be very different and have the same ideas. One day, I went to look for the video of the conference on the Internet: I saw a Peter Higgs who seemed both embarrassed to be the star of the day, eyes staring vacantly and applauding very slowly, as if absent, but also certainly very moved by this discovery. It must be said that between the prediction and the discovery of the particle that bears his name, 48 years had passed — an absolute record.

Peter Higgs and François Englert had confirmation of the validity of their hypothesis during their lifetime. Unfortunately, this was not the case for their colleague Robert Brout, who died in 2011. Moreover, his name was not selected for the Nobel Prize in Physics, which cannot be awarded posthumously, and which was therefore awarded only to Higgs and Englert in 2013, one year after the discovery of the boson. Cruel, isn't it?

Let's leave the auditorium. Why don't we get a little height? Follow me; we'll take this elevator and go up to the top floor, the sixth, to enjoy the view. We could have stopped at the fifth floor and knocked on the door of the General Manager's office, pretending it was a mistaken direction... but we're not going to play that game.

Let's go for a walk on this balcony instead. From this side, you can see a large part of the buildings of the Meyrin site, with the Jura as a backdrop.

CERN and Jura

But the scenery is even more beautiful on the other side. On the left, you recognize the Globe, of course.

Opposite, in the distance, you can see the Alps: the highest peak in the chain is Mont Blanc! On the right, you have, of course, recognized building 40, which is already familiar to you. Finally, below, you can see the terrace where we just had lunch, the shell of the dipole magnet in the middle of the lawn, as well as solar panels on the roof of the restaurant annex. I told you that we're ecological at CERN.

After contemplating this panorama, let's go back down to the first floor.

The Globe seen from the terrace

The landscape seen from the terrace

Visit of buildings 1, 4, 5, 52 and 53

I suggest we continue our visit by taking the corridor in front of the auditorium. I like this bay window; I find it very pleasant.

All the offices we pass on the right are occupied by CERN's press office. Be careful — this is not the CMS or ATLAS communication group, but rather the CERN communication group. I would remind you that CERN is responsible only for accelerators, in particular the LHC. The various experiments, such as ATLAS or CMS, are international collaborations indepen-

Corridor near the auditorium

dent of CERN: each has its own communication group. Isabelle Wingerter told us earlier that the ATLAS collaboration consists of 180 institutes in 38 countries, and represents about 5000 people. As for the CMS collaboration, it brings together about 3500 people from more than 200 institutes in 46 countries. From one year to the next, all these figures are likely to change; but in any case, it remains true that ATLAS and CMS are two of the largest scientific collaborations in the world!

We reach the red lockers and the large black Nespresso machine. It must be about 2 m high. One day, I was naive enough to think that it was free and struggled for a while with it, before noticing on the left side a little hidden place for inserting the coins.

Don't you find all these red lockers intriguing? At the beginning of my internship, I walked along them for a month without really paying attention to them, until I wondered what those 303 big red lockers were doing in this corridor. Yes, 303: one day when I had time to waste, I counted them. I'm not going to make the suspense last: they are archives of pre-publications sent

The red lockers

from all over the world to CERN in the days before everything was stored on the Internet. Apparently, there wasn't enough space to hold them in the library.

Hey! Why don't we open a locker, just for fun? Which one inspires you? This one, with the sibylline label: SU 65-01? All right, let's open it. It is filled with well referenced documents. Let's pick up an article at random. What is it about?

This is an archive from 1980, a physics seminar in Trondheim called "On the current status of

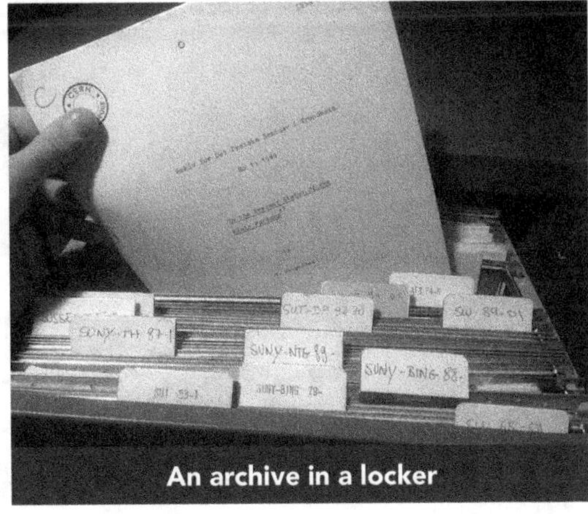
An archive in a locker

Klein's paradox." What a lucky draw, no? This article deals with a paradox of quantum physics, linked to the description of a particle confined behind a potential barrier... Okay, let's close the parenthesis, and the locker.

Turn right behind the red lockers. We are at the library: a real goldmine. My first experience here was quite successful: I was looking for a very interesting article by Paul Dirac, entitled *The Evolution of the Physicist's Picture of Nature*, so I came to this library and, just five minutes after entering it, I had the printed article in my hand. I can't even imagine everything we can access here, since the works on all these shelves and in these red lockers must be only a tip of the iceberg, if we consider all the digital archives invisible to us.

We will make a small detour by going down one floor by this staircase... In your opinion, why is the corridor in front of us exceptional?

Answer: That is where the Web was invented! We are in a real place of history, as shown by the commemorative plaque on the left wall. It points out that Tim Berners-Lee and Robert Cailliau, two CERN staff members assisted by students and post-doctoral fellows, developed the Web in these offices between 1991 and 1994. Tim Berners-Lee then went on to lead the W3C, the "WWW Consortium," a global organization funded in part by CERN, which would then complete

The Web Corridor

the World Wide Web project. But this is where it all started!

A legend tells us that the first Web servers were stored in a room on the fourth floor, number 404, and that if a request was not successful, the message "Room 404: file not found" appeared on the screen. So this would be the origin of the famous 404 error... Except that in this building, there is no fourth floor. So it must be a pure legend. Nevertheless, I was very disturbed the day I discovered, just before the entrance to this corridor — come and see — these two grey doors on the left, identified as R-404!

Pure coincidence? I like to think that this is where error 404 originally came from, even if there is a good chance that this explanation is false. I have never found anyone, even among the people who work in this corridor, who can solve the mystery associated with room 404...

After this brief moment of reflection in front of the place where the most significant invention of the end of the last millennium was born, I suggest going back to the library and crossing another corridor. Doesn't this one remind you of a movie, by any chance?

You're probably going to blame me for a certain narrowness in my film culture, but honestly, it's once again like being in *Titanic*. If you listen, you may hear Jack calling for help in one of the offices: he is handcuffed to a pipe, and the water is rising. But in fact, in a more prosaic

The room n°404

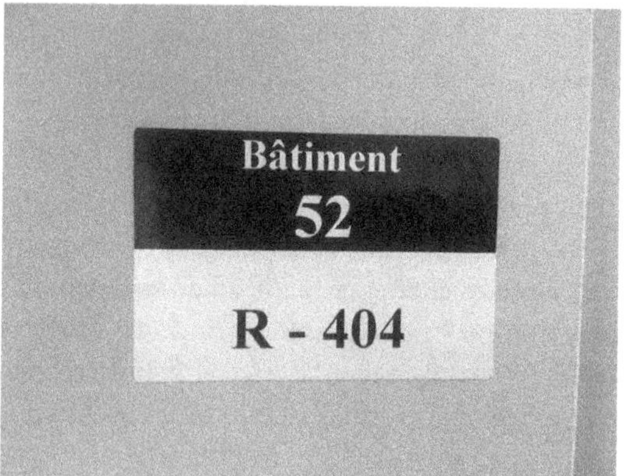

The "Titanic Corridor"

way, this corridor reminds me of my frantic dashes in order not to miss my bus.

Take heart — we still have one floor to climb. Soon, when we get to the end of the corridor, we will again be in the... Oh, my God! It's incredible. Have you seen this office? What a mess, it's amazing! Apocalyptic!

Who the hell has such an office? Wait, I'll look on the door... Oh, I understand better: it's the office of John Ellis, a theorist of Supersymmetry (or SUSY for regulars), and inventor of the analogy that associates the Higgs field with a ski slope. I criticize the apparent chaos but, in reality, I have no lessons to give him. In addition, this desk has a certain style that is proportional to its entropy — to its disorder, if you prefer. I believe that John Ellis is one of the few people on Earth who can legitimately keep his desk untidy and make it a trademark. He has a rare good excuse... Before we leave, do you see his blackboard?

"Inflation is a LIE." Brrrrrr, that's a bit scary... Let's get away before being flushed out by the owner of the place, whom I think I saw in the restaurant, with his beard and long white hair...

There is always a certain intellectual effervescence in this corridor. Look on our right: here are theorists who are struggling with difficult calculations... They seem very concentrated: let's not disturb them.

We finally reach the end of this long corridor... Do you recognize

John Ellis's office

John Ellis's blackboard

Two theorists in full discussion

it? Yes, we are back at the visitors' entrance, where we were early in the morning! This reminds me that there is a CERN TripAdvisor. No, it's not a joke! Like everyone on this kind of site, I was curious to see a negative opinion; of course, the vast majority of comments are excellent, but our scrupulous nature always leads us to read those from the minority of disgruntled people. Inevitably, I came across one huge mistake; a lady wrote that she regretted not having access to the detector in these terms: "It's like being on the *Champ de Mars* without being able to see the Eiffel Tower." We can understand her frustration, but we obviously can't go down to see the detector in operation in the same way we would visit the Eiffel Tower! There's radiation down there... It is necessary to explain to this charming lady that ionizing radiation can cause cancers or DNA mutations. By the same token, would anyone have the idea of writing a TripAdvisor comment such as: "I am very disappointed: we visited the nuclear power plant, but they didn't let us dive into the reactor pool!" So you ought to come at the right time, when the LHC is on break. But even during an operating period, there is nothing boring on the surface!

View of the Globe and the tramway

CERN clubs and cordiality of coffee

We are back next to the flags, in front of the tramway and the Globe. Not far from us is the CERN nursery school... I was very surprised when I learnt about it! And it is not small: it seems that there are nearly 200 children. You can really find everything here; in fact, CERN looks like a small town of 10,000 people.

I suggest we go to building 40 for coffee. But since the weather is fine, we're not going to retrace our steps inside; we're going to take this road over there, on the right.

Niels Bohr Road leading to building 40

A few steps away is the jazz club, where I enjoyed the piano during my internship months. Do you think that physicists are constantly immersed in their calculations, experiments, simulations or software? Far from it! CERN has a wide range of clubs of all kinds.

On the dedicated Internet page, I have listed 52 of them: basketball, cycling, volleyball, squash, football, table tennis, golf, hockey, swimming, canoeing, rugby, sailing, and above all, the ski club, which has a resounding success; this is the advantage of being close to the Jura and the Alps. If there are no more seats in the buses that leave early on Saturday morning for the slopes, you can always console yourself by going to the billiards club, the petanque club or the frisbee club. There are also musical activities: jazz club, dance club, vocal ensemble... And then there are the unusual clubs, such as the highly venerated Nature and Bees Club, which promotes beekeeping in a friendly atmosphere, the scuba diving club, or the softball club, which has two teams: the Leptons and the Quarks!

On our right, on the other side of the lawn, we can see the terrace of the restaurant where we had lunch

earlier. Let's continue along Niels Bohr Road: it leads us straight to building 40, which we can already see in the distance in all its majesty. That's where we're going to have a little coffee.

It's four o'clock — that's coffee time, isn't it? When I am at CERN, an internal clock reminds me at this time of day that I have to go to the cafeteria. But I don't think I'm the only one in this situation. I have the impression that researchers worship coffee. Statistics on a wall of the restaurant where we were testify to this: 5.3 tons of coffee were consumed in this single restaurant in one year! Physicists consume coffee trucks-full. In all conference meeting programmes, you will see that breaks are always referred to as "coffee breaks." We might as well have called them "tea time" or simply "break," but no: here, coffee is a must. It seems that physicists need constantly to reinvigorate themselves! Paul Erdös said that a mathematician is a machine that transforms coffee into a theorem. Here, we could say that the physicist is a machine that transforms coffee into an experiment or a theory!

As I see it, what I would call the "cordiality of coffee" has established itself in the research community. The principle is as follows: when a group of physicists you are part of ends up in the cafeteria, there are only two possible options: either someone pays for you, or you pay for everyone. I find this very user-friendly, I totally adhere to this doctrine. This avoids situations like: "Do you have twenty cents?" or "Wait, how much money do I have in my pocket right now?" It also prevents — it's a fact of life — the person who collects the money from ending up with a 50-cent coin in his hand and having to ask: "Well, who wants to take back the remaining 50 cents?" Then follows the traditional "No, for me it's ok, I didn't give enough to get it back." "Keep it, I don't care about it." Or "No, no, please, it's for you." But actually everyone wants the coin! Hence the importance of this principle of cordiality.

After this long walk along Niels Bohr Road, we are back in front of building 40, still impressive! Seen from here, it gives an idea of the size of the ATLAS and CMS detectors. If CMS were placed next to building 40, it would reach the third floor, while ATLAS would slightly exceed the height of the roof. But you will certainly be more aware of the size of ATLAS later, when we go down to its cavern.

Let's sit on the ATLAS side of the cafeteria. You can see that the tables are half on the CMS side, half on the ATLAS side. In reality, this separation does not make much sense, because we very often see ATLAS people settling on the CMS side,

Building 40 seen from the front

and vice versa. But in life, you have to make choices, and for our coffee, I choose the ATLAS side. Randomly, of course.

All right, let's sit down. Oh, look: two tables away from us is Peter Jenni. He is a former Spokesperson of ATLAS, a precedessor of Dave Charlton, who presented the news at the ATLAS Weekly meeting. Peter Jenni was in this position when ATLAS was built: he is a bit like the soul of the ATLAS collaboration. He is nicknamed the "founding father of ATLAS," a title which inspires respect! He always has a wide smile on his face; he is a very friendly man, whom I have often

seen in building 40. At the very beginning of my internship, I had confused him with the librarian, but fortunately he didn't notice it... Or maybe he was being very diplomatic.

A brief history of the search for the Higgs boson

I'm going to take advantage of this coffee break to go back to physics, and, more precisely, to the Higgs boson, because I would like to give you an overview of the history of the search for it. Indeed, I would not want you to suppose that only the ATLAS and CMS experiments at CERN were involved in the search for the Higgs boson, which dates back well before the 2000s! In fact, we were already looking for it in the 1970s. Why did it take so long to observe it? Simply because before the LHC, we didn't have an accelerator powerful enough to produce it. Let me explain.

In everyday life, you distinguish between energy and mass. When you see Novak Djokovic at Roland Garros returning all the balls from the back of the court, you say he has overwhelming energy, not that he is massive; that would even be a misnomer. On the other hand, Geluck's Cat, for example, is objectively very massive, but not very energetic: it is rather phlegmatic.

In contrast, in particle physics, mass and energy are equivalent, according to a formula that you cannot ignore: $E = mc^2$. In truth, this formula is valid only if particles are at rest; it is a little more complex when particles are in motion, but the idea remains the same. The idea is that energy can be transformed into mass, and that mass can be transformed into energy, with a proportionality factor of c^2. Note that c^2, the speed of light in a vacuum squared, is a gigantic number, 1 followed by 16 zeros, which implies that even a body of very small mass can contain gigantic energy.

What happens during a collision? When two protons meet head-on in the LHC, they release energy that depends on their mass and velocity. Then this energy "rematerializes," so to speak, into massive particles, still thanks to $E = mc^2$, or rather to $m = E/c^2$. But not just any of them! For in all these processes, there is a *sine qua non* condition, which I have already had occasion to mention: energy conservation, which is a fundamental law in physics. It requires that the total energy of the initial protons must be equal to the sum of all the energies of the particles resulting from the collision.

One of the accelerators involved in the search for the Higgs boson was the LEP in the 1990s: as I told you this morning, it was the accelerator that preceded the LHC in the same 27 km tunnel. Instead of sending protons against each other, however, it used to send electrons against positrons. Despite all the efforts made, the collision energy was not sufficient to create a Higgs boson; we learnt this after the fact, of course! Today, we know that the mass of the Higgs boson is about 125 GeV, the GeV being an energy unit adapted to the world of particles; to express it in terms of mass, you just have to divide by c^2. The mass of the Higgs boson is therefore about 125 GeV/c^2. However, at the time, the LEP only achieved 114 GeV/c^2.

The one that increased in energy, in the 2000s and until its closure in 2011, was the Tevatron in the United States; it was unlucky too. The Tevatron went fishing for new particles in the range [140 GeV–180 GeV]. It could hardly see events below 140 GeV, due to too much background noise. It was therefore unable to observe events involving a Higgs boson.

In 2009, the LHC produced its first collisions, and in July 2011, both the ATLAS and CMS experiments announced their first results: the Higgs boson was not in the energy range [150 GeV–200 GeV]. Finally, thanks to the LHC and the Tevatron, the noose tightened around the range [115 GeV–140 GeV].

You know what happened next: in July 2012, it was announced that a particle had indeed been detected at 125 GeV, with even more certainty since the independent observations of ATLAS and CMS were in agreement. We found it!

Here, this reminds me of a physicist's joke. A typical physicist's joke is — how can I explain it to you? It's a joke that is not bad enough to be forgotten forever (and that may even be worth repeating), but is not funny enough to make listeners laugh out loud. This definition can be condensed into "almost funny," or as "funny for a physicist." I'll give you an example: when CERN physicists started to observe on their histograms something that looked like a new particle with the properties of the Higgs boson, they called this "something" the Higgs-like boson — until the Higgs-like boson turned out to be the Higgs boson in July 2012. And at that time, you might have said: "The Higgs-like boson is no longer 'like', but we still like it."

You're beginning to smile: I can see that you have recognized a joke that perfectly meets the criteria of the physicist's joke. I do not want to show that physicists have no humour — quite the contrary — but simply that you have to get used to this kind of joke, and that the more you appreciate them, the more you are a physicist at heart.

I have another one just for you. Two atoms are conversing: one has four electrons, the other only three. He says, "Damn! I lost an electron." The other one says, "Are you sure?" And the one who has only three electrons answers, "I'm positive." You almost laughed, didn't you? No?

Well, in short, after intense efforts, technological progress finally led to the discovery of the Higgs boson, which reinforces the coherence of the Standard Model. This is both good and bad news: good, because the Standard Model is not called in question, and bad, because... the Standard Model is not called in question! Indeed, we can be delighted to have such a compelling theory, and at the same time, it does not explain everything, so we can regret that it is still not in contradiction with reality — something which would open the way to a new, potentially revolutionary physics.

In any case, we have made great progress in our understanding of mass. The Higgs boson is no longer an old hypothesis from 1964; it is a reality. In 2012, the real world told us: Yes! At 125 GeV, there is a particle that is responsible for the mass of others!

At this moment, you may be asking yourself a question that, for my part, tormented me a lot. From the beginning, I have explained that the Higgs boson is responsible for the mass of all the particles that have any. All right. But I also talked about the mass of the Higgs boson: 125 GeV/c²! You can feel the question coming: this boson gives mass to other particles, but... what gives it its own mass? It's a bit like the paradox of the village barber: the barber shaves the people in the village, but who shaves the barber? And the answer, in both cases, is... himself. The Higgs boson couples to itself to give itself a mass. I did not want to leave this problem unanswered, because I assure you that it can lead to unbearable insomnia. Prevention is better than cure — trust me.

It's time to rest our minds a little! And for that, there's nothing like a little meeting with someone who truly belongs here, Melissa Gaillard; you will soon understand why. We have to get back to the car again; her office is not very close. Let's go.

Let me summarize. Our protons, originally coming from the hydrogen bottle, have been accelerated successively in the LINAC, Booster, PS, SPS, and LHC. Then they have collided within the detectors and formed a myriad of new particles. And sometimes, the collision of two protons has resulted in the creation of a Higgs boson, which has itself disintegrated into other particles whose trajectory, velocity and energy can be measured.

But in practice, the only thing we observe is electrical signals. Actually reconstructing events, i.e., determining the particles and the collisions from which they originate from the electrical signals they have generated in the different parts of the detector, requires very thorough analysis!

Indeed, the reconstruction of events requires such computing power that hundreds of thousands of processors around the world are used. It is therefore natural to continue our visit at the CERN Computing Centre. And to tell us about it, who could be better placed than Melissa Gaillard, in charge of communication for the Worldwide LHC Computing Grid?

This time we go to the end of Weisskopf Road and take Feynman Road. By the way, as we drive: do

The glass corridor of the Computing Centre

you know what CERN's first computer was? His name was Wim Klein. Don't think that Wim Klein was the name given to a computer: he was a real person! Before computers arrived at CERN in the late 1950s, calculations were made from tables, the ancestors of calculators. As this task was long and tedious, CERN decided to hire Wim Klein to carry it out. At his job interview, the jury asked him to do a calculation, and Wim Klein gave a different result than the one obtained using the tables. But the prodigy was convinced that his calculation was correct, and suggested that the jury repeat its own calculation with the tables. After verification, the jury realized its error. It must be said that it is difficult to trap someone who can extract the 73rd root from a 500-digit number!

We are actually very close to the antimatter factory we already visited. Melissa's office is on the ground floor of this building, number 31. I'll park right next door. There is a pretty glass corridor in front of the entrance: we will have the opportunity to pass through it.

In the meantime, let's turn immediately to the left. We're a little early, but her door is ajar, she must already be waiting for us. Let's go in.

Interview with Mélissa Gaillard

The Worldwide LHC Computing Grid

"Hello, Melissa, thank you for welcoming us! We are eager to know everything about the Worldwide LHC Computing Grid.

- Hello to you! Please, have a seat. So you're visiting CERN?

Indeed, and we have just begun the part of the visit devoted to data analysis. So what is the Worldwide LHC Computing Grid?

- It is a set of computing centres located around the world that store, distribute and analyze LHC data. When the LHC was designed, it was quickly realized that it would be impossible to analyze all the data it would provide at CERN, hence the need to distribute computing power around the world. That's why the Grid was set up.

Could you describe its role in detail?

- In fact, the Grid is hierarchical. First of all, there is the level 0 computing centre, called Tier-0: it is the one based at CERN, located in building 513, just behind us..."

And that's very good news, because we're going to visit it very soon.

"... This centre has an extension in Budapest, Hungary. These two centres at CERN and Budapest represent 15 to 20% of the total power of the Grid. You probably know that in the detectors, the collected data are sorted: we only keep data we are interested in..."

Yes, you know it well: this is the trigger function, which I already told you about.

"... Well, when the output of the detectors is received, the selected data are sent to the level 0 centre of the Grid, where it is all stored permanently on magnetic tape. A copy of this data is stored and distributed among the thirteen level 1 centres, also known as Tier-1 centres: in case a fire or other tragic event occurs at CERN, the data are saved elsewhere. Then, the data are made available to physicists on request..."

Here is something quite surprising: in storage libraries, robotic arms come out to retrieve the required data on demand by grabbing the magnetic tape concerned, which the arm recognizes by means of a bar code!

"... Data is then analyzed, particularly in level 2 centres, called Tier-2s. Currently, there are about 140 of them in about 40 countries. The physicist receives the result of the data analysis he has requested, without knowing where the data has been stored and where it has been analyzed. Others ensure that the distribution of the calculation is optimised, in order to make the most of the Grid's performance by matching resources to needs. Moreover, even if the LHC is stopped and there are no more incoming data, the Grid continues to perform calculations, since physicists conduct data analyses continuously, even on December 31 at midnight! You can check this yourself by adding a line of code in Google Earth that allows you to see the activity of the Grid almost in real time.

But why is there a need for such computing power?

- Because there is a huge amount of data! For example, in 2016 alone, there were 49 petabytes of data recorded for the LHC alone, far beyond our expectations!"

It's gigantic: it represents more than 100 million hours of tape on a computer!

"Indeed, not to mention that, in addition to the actual data, the Grid also processes simulated data, right?

- Yes, indeed, there are also simulations, which represent a good half of the calculations. These data, obtained theoretically, are just as important as the actual data because they allow the results of the experiments to be compared with the predictions of the theory, which is then confirmed or invalidated.

Regarding the actual data, there is one thing I don't quite understand: the objective of data analysis in the Computing Grid is to reconstruct events, thus, among other things, to identify particles from collisions. But the trigger also does this, even before the Grid, since it must recognize the particles and reconstruct the collisions to eliminate those that are of little interest. So I wonder — sorry, a little naively: what remains to be done for the Grid?

- Well, to be precise, the Grid reconstructs in a much more exact way the events that have been selected, and which are by definition potentially interesting! The trigger recognizes and eliminates only well-known events, which are quite easily identifiable. A much finer reconstruction and analysis, made possible by the Grid, then becomes necessary.

This amount of data received by the Grid will grow very quickly in the coming years: do you know if any solutions are planned, so that the Grid can withstand the exponential data flow in the future?

- There are various options: providing additional computing centres, using cloud computing to outsource the calculation (with European projects such as Helix Nebula), improving processors... This will be necessary: with the High Luminosity LHC, the number of collisions will be multiplied by ten, so we expect to have at least ten times more data to process by 2025.

But since the Grid is ultimately made up only of traditional processors, wouldn't there be a way to use personal computers?

- It already exists! The LHC@home programme allows everyone to participate in research by providing the computing power of their computer when it is in standby. Together, the volunteers represent the equivalent of a level 2 computing centre on the Grid!"

All the same! Now let's ask for figures.

"If you don't mind, I'd like us to go into a little bit more detail about the statistics of the Grid, which are very impressive and worth mentioning! For example: you said that the Grid received 49 petabytes of data in 2016... But how many has it received since its creation?

- In 2016 alone, the LHC produced 49 petabytes of data, but the Grid distributed a much larger amount, for example 100 petabytes per month between July and October, which is huge. It is therefore very difficult to estimate how much data has been transmitted through the Grid. Since 2010, the LHC has produced 185 petabytes of raw data sent to the Grid. But to this should be added the data after analysis and the simulated data...

And, on average, what is the transfer rate within the Grid?

- More than 35 gigabytes per second, whether the LHC is running or not. In other words, every second, 35 gigabytes of data pass through the Grid, with peaks at 55.

It's unbelievable. And what is the transfer rate at the output of a detector, ATLAS for example?

- The transfer rate from ATLAS to Tier-0 is on average 2 gigabytes per second, with peaks at 7 gigabytes."

Confess that you can be jealous of such a high transfer rate! At this rate, you could transfer about thirty hours of music to your smartphone in one second...

"This is all very impressive. However, I know that many people wonder if these major projects are relevant to everyday life. So does the Grid have applications outside particle physics?

- Yes, indirectly. This Grid — which is the first computing grid to have been created on the basis of the 1999 work of Ian Foster and Carl Kesselman — processes only LHC data; but there are other computing grids in the world, technologically very much inspired by ours, that are used in other fields. They allow calculations and simulations to be carried out in astrophysics, the life and earth sciences, climatology, and medicine in order to improve research on the brain, DNA or rare diseases.

And finally: what is the French contribution to the Grid?

- In France there is a level 1 centre, the IN2P3 Tier-1 on the Doua campus in Lyon, as well as seven level 2 centres located in different laboratories, which I will quickly find on my computer… Here it is: the CPPM in Marseille, the LPSC in Grenoble, the LAPP in Annecy, the LPC in Clermont, the IPHC in Strasbourg, Subatech in Nantes and the GRIF — the Grid for Research in Ile-de-France — which includes several laboratories in the region."

Don't some lab names ring a bell? Remember, the CPPM, LPSC, LAPP and LPC are part of ATLAS France; we were talking about that this morning with Isabelle Wingerter. Everything overlaps!

"Now that you've introduced us to the grid, it's time to visit it… Thank you very much, Melissa, for this information, and see you soon!

- You're very welcome; have a nice visit of the Tier-0!"

I imagine that after such a discussion, which gave you a good overview of both the size and the computing power of the Worldwide LHC Computing Grid, you must be looking forward to visiting CERN's level 0 centre, the Tier-0. Let's go right now! As we said, it is in the building opposite. We will therefore have the pleasure of passing through this beauti-

ful glass corridor that connects the two buildings (ideal when it rains!).

Since the corridor is short, I just have time to give you an amusing statistic. With the power of computers in the 1960s, it would have taken 13.2 billion years to analyze 1 year of LHC data. In other words, with such computers, the calculations should have been started shortly after the Big Bang. Progress has therefore been made since then in terms of processor efficiency!

The inside of the glass corridor

The Computing Centre

Before entering the heart of the Computing Centre, I suggest we get a little higher: let's go up to the first floor. This will allow me to show you all the computers as seen from above, a sight which will certainly impress you, I am sure. It's at the end of this corridor, behind the window... There you go.

You are looking at the famous Tier-0! In fact, part of Tier-0, because Melissa told us that this centre has an extension in Budapest, with which it is directly linked by a 100 Gbits/s connection. In the room we are overlooking, which is 1400 m², there are precisely... Wait until I find my little paper... 121,766 processor cores.

Let's go down to see all these servers more closely; each of them contains about ten processors. We've arrived in the hall of the Computing Centre... Look at the screen in front of you.

The Computing Centre from above

The hall of the Computing Centre

Live activity of the Worldwide LHC Computing Grid

You have certainly recognized Google Earth: thanks to the plugin Melissa told us about, we can see almost live the activity of the Grid. The statistics at the top right of the screen show us the number of tasks being processed at this very moment, the number of processors in operation, and the transfer rate over the entire Grid: all these figures are constantly evolving.

Let's go into the Computer Centre; I'll speak a little louder, because ventilation systems make a lot of noise! In addition to the ventilation, there is a cooling system that maintains the ambient temperature at about 30°C. Do I hear you telling me that that is not very effective for a cooling system? My answer is that, if it were to stop, the temperature would increase by 3°C every 30 seconds. It would become a nightmare... At the end of this row, do you see what looks like a big grey chimney against the wall?

Cold air arrives through this conduit, then passes through the subfloor. It then emerges through grids like the ones we see in this cold alley.

Numerous calculators

A row of refrigerated servers

Servers dedicated to hosting CERN websites

The panel above indicates that the servers in this row — like most in this room — are dedicated to computing for the Grid. But this is not the case for all of them: for example, the servers in this other row are dedicated to hosting websites domiciled at CERN: there are about 15,000 of them!

Others are intended to provide all the services related to the network on the Meyrin site: WiFi, mobile Internet, telephone network, emails... Look at the big red pipes: these servers are so important that they have a cooling system that works even if there is a power outage. Indeed, in case of an emergency, it is necessary to be able to communicate with the outside in all circumstances!

We will now go down to the basement. What could there be in the basement? Remember what Melissa told us: the data that comes from the detectors, even before it is analyzed here or in the other computing centres of the Grid, is recorded (after a first processing) on magnetic tapes at Tier-0. So, to sum up: the calculation is done on the ground floor and the raw data storage in the basement. The Tier-1 and Tier-2 centres of the Grid allow additional backups and analyses to be performed.

Let's take the elevator to the lower level.

There is immediately less noise, it is more relaxing, and it is less hot. As we leave this room, I would like to point out that it takes a lot of cables to transport data from the various detectors to here. Lots and lots of cables... To be honest, there are 35,000 km of optical fiber. Can we even imagine what this means? It takes a quantity of cable almost sufficient to encircle the Earth to carry data from the detectors to the Computing Centre!

Here we are at level −1: I will immediately turn off the light, we will see the small robots better. (I am actually talking about the robotic arms that make it possible to retrieve magnetic tapes at the request of physicists.) Let's go to these incredible "magnetic tapes libraries," right in front of us.

The libraries in this room contain about 23,000 magnetic tapes, which are replaced every three years: the data is then duplicated on new tapes with a higher density, i.e., tapes that can record more information. Today, the majority of them contain 8 terabytes of data, and the best ones 12 terabytes, while the previous ones used to contain 1 terabyte. That's why we recently went from 50,000 magnetic tapes to 23,000!

Through the glass of this library — I don't know what other term I could use — you can see on the walls hundreds of cassettes containing the tapes. On the side of each is a bar code, which allows the robotic arms to recognize them.

Talking about robots, I think there's one in the back... Be careful, get ready, it is rather fast when it moves... And there it is, with its small light! It can carry up to three cassettes at the same time.

I find it rather fun to watch. Each movement of the robot means that a physicist somewhere in the world has made a request to get very specific data... In all, some 10,000 physicists working on LHC data have this privilege.

I hope you enjoyed this little entertainment! This is how our visit to the Tier-0 ends. We will now take this staircase that leads us directly to the glass corridor. It's time to get back in the car once again.

Magnetic tapes libraries

The robot picking up the data

Interview with David Rousseau

Machine Learning and particle physics

Don't worry, we will soon visit the ATLAS detector to conclude this day in style. But first, let's return to building 40 one last time to discuss a final subject related to information technology which is very popular at the moment in a large number of disciplines (including particle physics), and which will perfectly complement the visit to the Computing Centre: it is Machine Learning. To tell us about it, we will meet David Rousseau, who is a specialist in the integration of artificial intelligence techniques in scientific research, especially in the ATLAS collaboration.

The potential of Machine Learning in particle physics is both very varied and very promising, since such methods could be applied, among other things, to compute the simulation of events faster, to improve the determination of particle energies in the calorimeter, or to make the trigger even more efficient.

We are already back in the parking lot of building 40. We have an

David Rousseau in the cafeteria

appointment with David in the cafeteria. I know him well, because he supervised me when I did a short Machine Learning internship in Orsay in the ATLAS group of the LAL ("Laboratoire de l'Accélérateur Linéaire"), before arriving at CERN. All this brings back memories... Oh, isn't he the one waving at us?

Yes, that's him, let's go join him.

"Hello, David, it's a pleasure to see you here! We are back from the Tier-0, and after this visit, I thought it would be interesting to hear you talking about Machine Learning.

- That is indeed a subject I know something about! Please, take a seat.

Before getting to the heart of the matter, could you explain to us what Machine Learning is in general?

- All right. Machine Learning is an aspect of artificial intelligence. Until the early 1990s, artificial intelligence was seen as the writing of computer programs with fixed rules established by humans. But we finally realized that this did not work in complex cases: there would be too many rules to write by hand, and the effectiveness of programs would remain limited. The revolution came with automatic learning. The idea is as follows: rather than having a profusion of fixed rules, we define learning rules. In other words, we teach the computer to formulate rules itself.

But how does the computer determine these rules?

- It is given learning examples, from which it induces rules. For instance, we want the computer to create an algorithm to distinguish dog images from cat images. It is therefore given a large number of dog and cat images, specifying each time which animal it is. Little by little, by comparing the images with each other, the computer will learn to distinguish a cat from a dog, and will define rules that will subsequently allow it to determine the animal shown in an unknown photo.

What was the first application of Machine Learning?

- It was the reading of postal codes, in the 1990s, to sort the mail! To do this, the computer is shown handwritten numbers from 0 to 9 so that it can learn to recognize them, and then identify the postal code written on an unknown letter. In the 2000s, Machine Learning developed considerably with the arrival of Google and Facebook, among others.

What applications has enabled the rise of Machine Learning?

- There are three that come to mind. First of all, there are automatic translations. Before, rules for translating from one language to another were written by hand, often with inconclusive results. Now, with Machine Learning, we give the computer a lot of French texts with their English translation to make it learn expressions and the different contexts in which a word can be found. Today, translations are much better.

There is still room for improvement...

- Absolutely! A second application is the interpretation of speech, that is to say, the written transcription of what someone says. Some computers, for example, have a dictation tool.

And advertising on the Internet... Doesn't it use Machine Learning?

- Yes, that is the third application I was thinking of! We are talking about targeted advertising. Depending on the webpages visited, Google is able to recognize your interests, and then shows you sponsored links that are supposed to match your tastes. Facebook can also determine your interests by identifying the content of your photos, for example. Then, how Facebook presents content to you will depend on what is likely to be of interest to you.

We feel like we're being tracked... But let's not get into the endless debate on "Facebook and privacy," and let's come to science: how can Machine Learning be used for particle physics?

- I was just talking about classifying dog and cat images... In fact, in particle physics, we classify images all the time! Roughly one billion events are recorded per year, and we try to classify them to separate those that resemble a particle of interest (a Higgs boson, a supersymmetric particle...) from those that do not.

How do we do this?

- The principle is always the same: the computer, from simulated events provided to it, learns to distinguish signal events (with the Higgs boson, for example) from other background noise events. It then establishes rules integrated into an algorithm capable of sorting unknown events, thus minimizing statistical error. Currently, Machine Learning research also involves finding methods for sorting algorithms to minimize systematic errors.

What do these errors mean?

- Statistical error is related to the number of measurements: the more measurements there are, the more the statistical error decreases. Ideally, if we had an infinite number of events, it would be nil. As for systematic errors, they are of two types: known unknowns and unknown unknowns. The known unknowns are the uncertainties regarding theoretical values determined by the calculation, or the uncertainties related to the calibration of the detector, which cannot be absolutely perfect. The computer is informed of these possible sources of systematic errors so that it can optimize the signal-to-noise separation by taking these errors into account.

But it is not the known unknowns that are the most frightening...

- No, the most frightening ones are the unknown unknowns, of which, by definition, the Machine Learning algorithm cannot be informed! These are the things that we would not have thought of... A good example is the OPERA experiment, which measured neutrinos going faster than the speed of light, due to a poorly connected cable... To avoid this type of error, there are two generalist experiments at CERN with the same research fields: ATLAS and CMS. In addition, internally, there are many verification processes in place to ensure the reliability of the results.

Let's talk about the future... The application of Machine Learning to particle physics is very recent, and I imagine that it opens many perspectives?

- Indeed. For example, the trigger system uses practically no Machine Learning techniques, whereas its role is to classify! The idea is therefore to integrate Machine Learning into the trigger in order to improve the identification of potentially interesting events. Here is another perspective: research is being conducted to use Machine Learning to improve the determination of the energies of particles passing through the calorimeter. All things considered, the distribution of the energies deposited in the calorimeter at a given moment is like an image to which Machine Learning methods can be applied...

Finally, let's talk a little bit about you: what is your personal contribution to the ATLAS experiment?

- In the 2000s, I worked a lot on the development of classical computer software for the interpretation of ATLAS data. I was responsible for all ATLAS software (excluding the trigger) that starts from raw data and reconstructs events.

Do these data analyses require the Computing Grid?

- Of course, because you need a lot of computing power! From 2011, following my meeting with another researcher from my laboratory who was already doing Machine Learning, I started on this path. Today I am ATLAS–Machine Learning coordinator.

And I guess you'll be here for a long time! One last question... Without holding back, what do you think of relations with CMS? Isn't there a little rivalry between you two?

- Yes, but without any low blows; everything is going well! I get along very well with the people at CMS, with whom I have very fruitful exchanges. Our common objective is to develop useful tools for particle physics for all experiments, even if in practice we work for one of them.

Thank you very much, David, for all these explanations!

- See you soon, and all the best.

Thank you."

I believe that this interview makes a good conclusion to the part of our visit devoted to data analysis, because it opens up perspectives for tomorrow's computing in particle physics, with an increasing contribution from Machine Learning.

We are now entering the last part of the day and approaching the crucial moment: the visit of the ATLAS detector, the largest particle detector ever built! But I have enough time to keep the suspense going a little longer. You can well imagine that a detector like the one whose operation I outlined to you this morning requires special 24-hour surveillance. That is why, in the control room we will be passing through, people called "shifters" are constantly verifying that there is no problem with the detector: it is a rather tedious but essential task. There is a simple rule: each laboratory involved in the ATLAS experiment must devote one third of its time to community work; in other words, each member of a laboratory must give on average one third of his time to the community, in particular by performing shifts.

Before leaving for ATLAS, I suggest you meet a shifter — Alvaro, whom I see two tables away from us. Alvaro is doing a PhD, and his particularity is that he looks like a Viking, while he is of Spanish origin. Let's go see him.

Shifts on the ATLAS detector

"Hello Alvaro! I guess you must have finished your shift not so long ago?

- Hi! Indeed, I've just come back from the ATLAS control room. I've been monitoring the state of the calorimeters for the last eight hours.

Eight hours! Without interruption?

- In principle, yes, which makes three relays per day. At my workstation, I have three computer screens in front of me that give me live information related to calorimeters, such as temperatures and electrical voltages. During my period of surveillance, I make sure that there is no temperature rise, voltage surge, or leak somewhere.

So you're in charge of all the calorimeters?

- Yes, the electromagnetic calorimeter in the centre and its caps on the sides, as well as the hadronic calorimeter.

What if there's a problem?

- If there is any problem, I see it on the screen: on the diagram of the detector, the part that has a problem is indicated by a color code. If it is yellow, it is a minor problem, but it must be fixed immediately to avoid a possible aggravation. For example, the best temperature for liquid argon in the calorimeter is 88 K: if this temperature increases to 89 K, the part concerned turns yellow on the screen. If it's orange, the problem is getting really serious; and if it's red, it's fatal. Yellows are frequent — there are some every day; oranges are much rarer. As for the reds, in about 200 hours of shift time, I've never had any. Lucky me!

200 hours?! It is true that it takes people to monitor 24 hours a day: whether for technical stops, tests or data taking; the work on the detector is unceasing. But haven't you been bored all this time?

- Not really. In principle, shifters should monitor the parts of the detector under their responsibility without doing anything else. If you take a coffee break, for example, you have to do it at the proper moment so you don't leave when things are bad: so you just have time to get your coffee and come back! For my part, I only did day shifts.

Ah, why didn't you do a night shift? Some people must envy you!

- Because I got my shifts on my computer two minutes after the reservations opened: first come, first served! I didn't want the night shift, because it's not my rhythm. But, even if it is less popular, some people prefer the night because there are statistically fewer problems to manage.

I imagine there must be many shifters in the control room. How is all this organized?

- There are always 7 shifters. Some people check the status of a part of the detector in real time: there is a shifter for the pixel detector, one for the calorimeters (my job earlier), one for the muon detector, and one for the trigger, who ensures that the right amount of data is collected to avoid saturating the detector. There is also a shifter in charge of the security of the staff, the detector and the cavern containing it; last Friday, for example, he was the one who had to deal with a water leak in the cavern, next to ATLAS..."

He is also called SLIMOS, Shift Leader In Matter Of Safety.

"... There is also the Run Controller, who configures ATLAS, controls all the machine parameters and modifies them according to what you want to study. Sometimes you acquire real data, sometimes you do tests — that is decided every day at a coordination meeting. And finally, there is the Shift Leader, who centralizes all the useful information from the ATLAS shifters and the LHC, which provides the beam. In the event of a problem, he is the one empowered to take decisions, after consultation."

It seems to me to be a well-oiled mechanism, no? This requires a certain amount of manpower to keep the detector in good working order!

"And what should you do in case of a problem?

- I have to call an expert, who has to be within a half-hour's drive of the control room. No ATLAS shifter is there to solve problems on his own. It is the expert who will decide on the procedure to follow; and if the expert does not know what to do himself, he calls on a "super-expert," who knows the machine like the back of his hand. But this situation occurs very rarely because most of the problems that occur, such as voltage or temperature peaks, are well known. Then, it is the Shift Leader who decides whether or not to stop the data collection, or simply the failed part of the detector if it does not affect the others. For my part, with the calorimeters, I was rather at ease: the pixel detector and the muon detector have more problems these days."

So, for those who want the least trouble possible, we can conclude that the best plan is to be a calorimeter shifter at night!

"Alvaro, a last question: you only told us about the ATLAS shifters, who deal with a detector, but there are also the LHC shifters, for example, who work in the accelerator control room. Is their work different?"

- Hmm I can't say too much because I've never done an LHC shift, and anyway it wouldn't be possible. To be

a shifter at the LHC, you already have to be an expert, because the stakes are higher: these shifters provide a beam for all the experiments, so if something goes wrong in the tunnel, all the experiments are penalized! To be here, by contrast, for ATLAS physicists a three-day training course is enough. As for their work, I imagine it is similar: undoutedly, they must check the proper functioning of the magnets in the tunnel and the cooling systems.

Well, thank you, Alvaro, for sharing your experience with us!

- It was a pleasure!"

Now that Alvaro has put us in the atmosphere of the control room, all we have to do is to go there. Once again, let's go find the car. I'll let you say goodbye to the beautiful building 40, which *a priori* we will not be seeing again today.

We leave CERN through the entrance next to the jazz club, where we not so long ago. We pass again in front of the Globe, in front of the flags, and... this time we turn right, towards the ATLAS control room. There is, of course, a badge check: don't forget to take out your visitor's card again. It is much easier to park here than in front at the parking lot for visitors; it is the privilege of being an ATLAS physicist, hence a VIP. VIP, like the Higgs boson. That's right: a VIP is a *Very Important Particle*. (I couldn't help saying that, forgive me. It will add an additional physicist joke to my credit.)

The control room is adjacent to the building with the elevator leading down to the ATLAS cavern. As soon as we get out of the car, we find ourselves facing this splendid mural fresco next to ATLAS.

Do you think this fresco is life-size? No: it's only on a scale of 1/2! Once again, this gives an idea of the gigantic size of the detector, since this fresco is "only" 12 m high. I find it remarkable: cheerful, colourful, and appropriate! It allows you to see the detector both from the side and from the front: it's clever.

The ATLAS mural fresco

The ATLAS control room

Continue straight ahead until you reach the gate. You need to use your badge again to pass: the ATLAS control room is not easily accessible. I'll let you fight with the turnstile. This sign "it's forbidden to play football in the turnstile" is surprising. What a pity — I had just brought a ball...

We arrive in the public area: welcome to the ATLAS "Visitor Centre"! The control room is located just in front of us, behind the large window.

Pictograms on the turnstile

You will admit that this place really looks like an aquarium: you could stay for hours in front of this glass wall and watch the shifters in front of their computers.

The ATLAS Visitor Centre

The ATLAS control room

Impressive, isn't it? On the right wall eight images are projected on a large screen, each corresponding to a part of the detector, with diagrams, graphs, and numbers that represent corresponding information in real time.

Did you notice the little camera on the wall next to the TV? It is actually a webcam, because the ATLAS "Virtual Visits" take place here. For an hour, a guide introduces his audience to the ATLAS experiment and particle physics, through an interposed screen, with the control room behind him. Students from all over the world can enjoy these presentations remotely, and can ask their guide any questions they want.

Finally, a last little curiosity in this antechamber of the control room: the ATLAS detector... in LEGO! It is

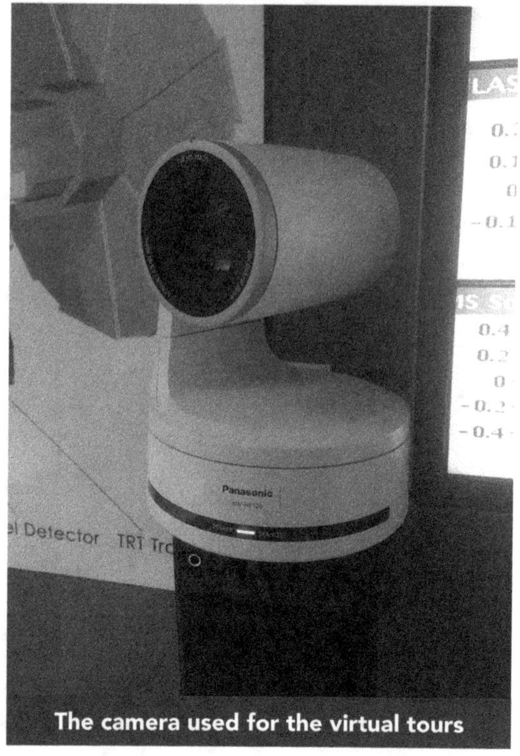

The camera used for the virtual tours

important to note that we do not see such sophisticated LEGO constructions every day: an ATLAS physicist assembled nearly 9500 pieces!

You've been waiting for it, and here it is: now, the real adventure begins, the one in the heart of the ATLAS cavern, the one in the heart of the Earth... Let's leave the Visitor Centre. We only have a few steps to take...

The ATLAS detector

We arrive in front of the entrance of the SDX 13178 building: the ATLAS detector is practically under our feet, 100 m underground! We are close to the goal.

We are approaching the elevator that will take us on a journey to the centre of the Earth. But first, there is this airlock...

Can you see the small black box on the right through the glass of this green door? It is a biometric reader. At CERN, retinal scans that open

The entrance to the SDX 13178 building

doors are not science fiction! I'll go through alone; I'll open the visitors' airlock next door for you, and we'll meet on the other side.

The airlock

There we are, near the elevator; let's start by taking one of the helmets hanging on the wall.

I'm calling the elevator... here it is already.

It is indeed a big elevator! Don't worry: if it breaks down, there's another one. You can press "–2". We start the descent, very gently...

This elevator is not like others. We've all had the opportunity at one time or another in our lives to take elevators that go very high. But taking an elevator that goes down very deep is much less frequent! And certainly, I admit, much more claustrophobic. But what wouldn't we do for science?

As we descend into the bowels of the Earth, you might legitimately wonder why the accelerator was built an average of 100 m underground. Why not shallower, and, above all, why not on the surface? There are environmental and economic reasons: actually, building 27 km of infrastructure on the surface is more expensive (especially because of the price of land!) and less environmentally friendly than digging a tunnel — all the more so because the neighbourhood of the region of Gex might not have been delighted. Imagine yourself saying to a local resident, "Excuse me, but we're going to uproot the hydrangeas in your garden, because we're going to put in a particle accelerator instead..."

In addition, this average depth of 100 m was chosen because it corresponds to a geological layer of molasse that is easy to excavate. As this is inclined towards Lake Geneva, the LHC itself is inclined by a few degrees, in order to remain within this layer. This is why the depth of

The helmets

The elevator to the ATLAS cavern

the LHC varies between 70 and 140 m, depending also on variations in the surface terrain.

But there is another even more subtle reason: during the visit of the AMS, in particular, we mentioned cosmic rays, those flows of very energetic particles that come to us from space. Placing the detectors at some depth underground considerably reduces these cosmic rays. Those that nevertheless reach the detectors are used to perform calibration, that is, to check the proper functioning of the detector and test its performance when it is traversed by very high energy particles.

Here we are: a little maze of corridors begins here... Let's walk to the red door.

Another restricted access door, you might say. But you don't enter the ATLAS detector cavern as you would any ordinary place; the same goes for the other detectors and many other installations at CERN, by the way. Let's go on.

A door with restricted access

The corridor to the ATLAS cavern

We're getting to the last corridor! Behind the yellow door at the back, ATLAS awaits us. The suspense has lasted too long, I admit. But I consider that concluding a CERN visit with the ATLAS detector is an apotheosis.

I'll open the door for you, and let you go first. Beware of the shock, the amazement... And... TADA!

I would like to introduce you to the ATLAS detector — in the flesh, so to speak! It's really unbelievable. No matter how much we talk to you about the gigantic size of the detector, 46 m long and 25 m in diameter, we can fully appreicate it only by seeing it with our own eyes. We can add to this the fact that ATLAS weighs 7000 tons, which is equivalent to the weight of the Eiffel Tower.

Forgive me, but I'm always rendered speechless when I stand in front of ATLAS. Beyond the technological prowess represented in front of us, I believe that such an achievement can be considered as a work of art. It's simply beautiful. I remember Dave Charlton, the

Part of the ATLAS detector

current ATLAS Spokesperson, saying in a TV show, as he stood next to the detector, "We really built this thing?!" We can only share this stupefaction!

In reality, we are seeing only a small part of the detector, one end of the cylinder. Since the ATLAS sub-detectors are nested inside each other, like Russian dolls, we can see only the one on the outside, namely the muon detector. It consists essentially of a myriad of rectangular muon chambers, the total area of which represents more than three football fields. We can clearly see one of these chambers just above us.

Behind the scaffolding is a kind of large pipe with orange bands. It is one of the eight coils of the magnet that creates a toroidal field in the muon detector and gives a helical trajectory to the muons. However, since the temperature within these magnets is close to absolute zero, water is likely to form on the surface. For example, take a cold can of beer from your fridge: the outside will probably be damp. Under these conditions, water droplets could leak onto the nearby sub-detector below. This is why an electric current flows through the orange bands to evaporate the water.

A muon chamber

Now look at this big golden wheel on the right: it is one of the two muon wheels at the ends of the detector, whose role is to detect (naturally enough) muons.

This wheel is the largest component of ATLAS, and here is what's really amazing: muons are detected with an accuracy equivalent to the thickness of a hair! In other words, the position of this huge wheel in relation to the rest of the detector is known to about one-tenth of a millimetre. You are obviously amazed, and I understand your reaction perfectly.

While we admire ATLAS in all its greatness, in every sense of the word, I would like to talk briefly about its construction — an absolutely titanic undertaking. Before even assembling the detector, it was necessary to dig its cavern; as you can see for yourself, its volume is considerable. No less than 300,000 tons of rock had to be extracted! It is customary to say that this cavern could contain the nave of Paris Cathedral, which represents 11 floors of walkways. The excavation of the cavern began in June 1999, and over the next two years, thousands of tons of excavated material had to be brought to the surface. Above our heads, we can see the large hole through which the various components of the detector were lowered.

After pouring concrete to a thickness of 5 m on the cavern floor so that it would not give way under the weight of the detector, the feet of ATLAS were installed. Then, over

One of the two muon wheels

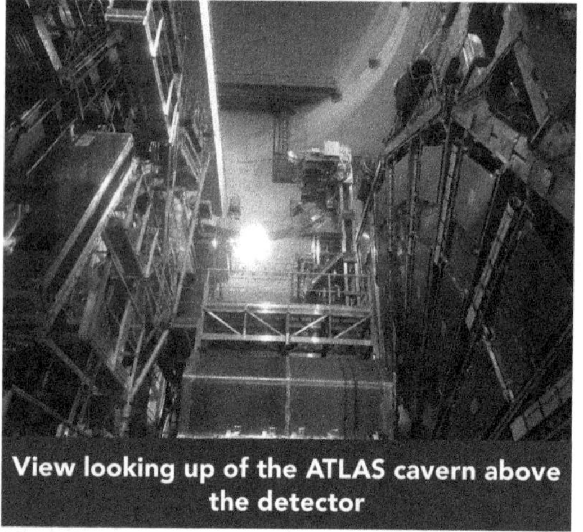

View looking up of the ATLAS cavern above the detector

the years, the detector was assembled in the cavern in pieces, in no particular order: the solenoid magnet, the hadronic calorimeter, the coils of the toroidal magnet, the electromagnetic calorimeter, the pixel detector in the centre and its 80 million reading channels, the muon wheels — not to mention the 3000 km of cables! The construction was completed in March 2008: it took almost 10 years to build ATLAS!

The ATLAS detector requires continuous maintenance: physicists must regularly realign it, i.e., check with extreme precision that the sub-detectors are in place relative to each other. The calibration of the different parts of the detector must also be repeated. In addition, the components wear out and need to be replaced. ATLAS will thus undergo improvements, of which Isabelle Wingerter gave us an overview this morning. In short, given the possi-

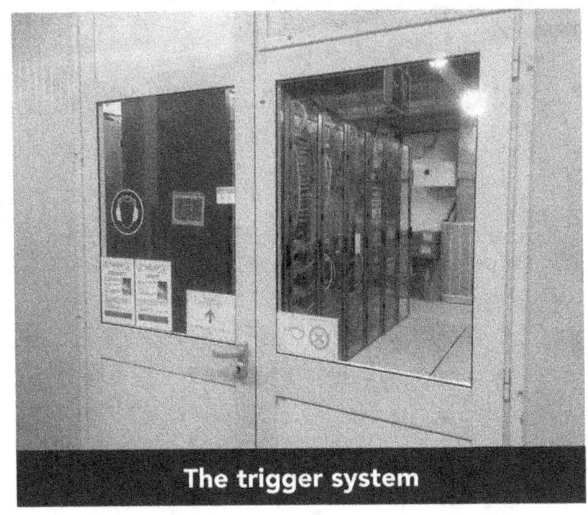

bility of choosing ATLAS, I am convinced that the Danaïdes would still prefer their sieve.

It is amazing to see what resources are used to probe the smallest constituents of matter. I think that, faced with such a mastodon, you can get a sense of all the technological prowess that had to be deployed to build it. But that's not all: bear in mind that this huge detector of unequalled complexity was developed to prove the existence of the Higgs boson, that is, to verify the prediction made in Peter Higgs's article, which is only a page and a half long! The letter of intent that marks the birth of the ATLAS project dates back to 1992; the Higgs boson was discovered in 2012. It took 20 years of intense effort to validate a page and a half of theory. Isn't that fantastic? Science advances with giant strides, and we do well to be amazed by it!

Before going back to the surface, I suggest you take a look through this door, which we will not be able to pass through. Can you guess what it is? Here's a hint for you: I told you about it this morning...

It's the trigger system! Remember: this is the system that sorts events, keeping only the very small fraction of them that is potentially interesting. In fact, the first instance of triggering — the one that performs a first rough sorting — is located on the detector itself. More precisely, each part of the detector has its own sub-trigger, which sends the selected events to the next trigger, which is in front of us. This second level trigger gathers data from all sub-triggers attached to the detector, in order to reconstruct events and perform a second, finer sorting.

Prospects for the future

I think it's time to go back to the surface, and soon time for you to leave us, after this busy day! Nevertheless, before you leave, I still have three important things to tell you. The elevator is waiting for us: let's go. The large doors close, and we begin our ascent to the ground floor...

First of all, I obviously need to talk to you about the prospects for the LHC. As we have had several opportunities to mention today, the LHC will undergo an upgrade phase in 2024–2025 to become the High-Luminosity LHC, or High-Lumi LHC, which will provide about ten times more collisions than currently. Since research in particle physics focuses on rare events, having more collisions — and therefore more data — is essential. Although the LHC was built primarily to discover the Higgs boson, even after this objective had already been achieved, between 2010 and 2016 it produced only 5% of the total number of collisions expected prior to 2037. So we will probably have many further surprises in the future!

In parallel, all other accelerators upstream of the LHC — including the Booster, PS and SPS — will also be upgraded to provide the required beam for the High-Lumi LHC. In addition, LINAC 2 will be replaced by a more powerful linear accelerator, LINAC 4. Where is LINAC 3, do you ask? In fact, as I told you this morning, LINAC 3 already exists and serves as an injector for the LEIR, the lead ion accelerator.

We have arrived at the ground floor, don't forget to put your helmet back on the wall. I'll open the doors of the airlock for you; I'll pass through the side door, as I did before. We will take the car one last time to return to the front of the Globe: that's where it all starts, and it all ends!

Both High-Lumi LHC and LIU (LHC Injectors Upgrade) projects are "short-term," if I may say so, and are underway. But we can already look further into the future: hypothetical projects for even more Pharaonic accelerators are being studied. Overall, two questions will have to be settled.

The first question: after the LHC, will we continue to produce hadron collisions (like protons), or will we produce lepton collisions (like electrons and positrons), as in the past? Debates are ongoing...

The second question to be decided is closely linked to the first: will a new 100 km ring be built under the Jura and Lake Geneva, or a 30 km-long linear accelerator? The circular accelerator project, called FCC (Future Circular Collider), would be specially designed for proton collisions, which could be raised to very, very high energy. Electrons, which lose a lot of energy when they follow a curved trajectory because of their intense synchrotron radiation, ideally require a linear accelerator, such as the ILC (International Linear Collider). Compared to protons, electrons do not allow us to reach very high energy ranges, but they could be effective for studying in detail the one in which the Higgs boson is located, which is far from having given up all its secrets. It is even planned, despite the energy losses associated with synchrotron radiation, to create electron collisions with the 100 km circular accelerator and then to recycle the latter by replacing the electrons with protons... Many options are therefore possible, but, given the costs, not all of them are likely to be implemented.

We're back at the Globe! I'll park in the lot in front of the main entrance; then we will head for the tramway, which is waiting for you.

The illuminated Globe

So that's the main thrust of CERN's prospects. Now, there is something else I would like to tell you: I want to share with you what I found really extraordinary at CERN. Today, during our interviews, we heard some cries from the heart: it is now time for mine. This is my analysis, which I could only formulate at the end of my internship.

There is what I would call a "global complexity" and a "local complexity" at CERN. By local complexity, I mean that each accelerator, each part of a detector such as ATLAS, and every step in the analysis chain, requires incredible technical and intellectual resources. Think of all the interviews we had today with experts in such diverse fields! I hope that through these meetings, and more generally through the entire visit, you have gained some insight into the enormous challenges that face each working group on a daily basis.

But this local complexity is seen to be coupled with a global complexity, when we take a step back and consider the entire experimental research process from the hydrogen bottle to the announcement of a discovery. And all stages in this process are interdependent! Consider, for example, the following simplified chain: hydrogen bottle, LINAC 2, Booster, PS, SPS, LHC, ATLAS, data storage, data analysis with the Worldwide LHC Computing Grid. If one of these links fails, the entire chain is affected. It has already happened: for example, between 2008 and 2009, the LHC was shut down due to a faulty junction between two magnets that caused significant material damage. Even more unexpectedly, in April 2016, the LHC did not work for a few days because of a stone marten that created an electric arc in a transformer. Let's have an emotional thought for this poor little animal, which, unfortunately for it, did not die for but "against" science. In short, research at CERN sometimes seems to hang by a thread.

And yet, despite the complexity of each task taken individually (local complexity), and despite the large number of tasks that must coordinate perfectly (global complexity), the Higgs boson, for example, was observed very quickly, and, what is more, by two detectors with different technologies! In my opinion, we can only really understand the exceptional nature of this discovery by being aware of everything that conditioned it: this was precisely one of the objectives of our visit.

I could have ended with this thought, but I have one last thing to add. I would like to point out that this day would not have had the same charm without the participation of all the people we interviewed, who shared their experiences with us. So let's thank them again warmly! In order, let's thank Bernard Pellequer, who introduced us to CERN with passion and encouraged me to propose this visit, but also Arnaud Marsollier, who meticulously explained to us how CERN is useful to us all, and why such an organization deserves our support. Let's thank Elias Métral, who unveiled the mysteries of particle acceleration, and who was particularly involved behind the scenes, but also Django Manglunki, who opened the doors of the accelerator control room to us and shared some of its intimate aspects. Let's thank Isabelle Wingerter, who introduced us to ATLAS France and detailed the contributions of the French laboratories concerned, but also Laurent Serin, who took us to the heart of the test beam hangar and gave us an example of an improvement planned for the ATLAS detector. Let's thank Melissa Gaillard, who told us about the Worldwide LHC Computing Grid and revealed its scope, but also David Rousseau, who explained the applications of Machine Learning in particle physics, and without whom — I will not hide it from you — this visit would probably never have taken place. Finally, let's thank Alvaro Lopez Solis, who told us about his shifts in the ATLAS control room.

And that's not all: some people also worked backstage. Many thanks to Marumi Kado and Claire Adam-Bourdarios, who also contributed to the smooth running of this visit!

But I think your tramway will be starting soon. I was very happy to share my experience with you: many thanks to you for accompanying me through CERN's entrails! I don't know what you will take away with you from this excursion, but if you are going to retain only one impression of this day, I would like you to remember a wonderful collaboration that transcends political, religious and geographical boundaries, and takes advantage of the mixing of cultures to advance human knowledge around unifying values such as commitment to science, professionalism, integrity and, especially, respect for others. I believe that in today's unstable world, where Europe is criticized on all sides, CERN is a remarkable example of international understanding and a model of living together.

Wherever you come from, I wish you a safe trip back — although, after such a journey into the subatomic world, in which some of the greatest mysteries of the universe lie, I don't know if the most difficult thing for you will be to go back home, or to come down to Earth!

Acknowledgements

The production of this book (the French version) was made possible thanks to a crowdfunding campaign on the *Ulule* online platform. I would like to very warmly thank all the generous donors without whom this book would not have been possible, including Elie Aslanides, Baptiste Barreau and Camille Raffin, Emmanuel Berthier, Elian Bouquerel, Jean-Marc Chebat, Yvon Choi, Jérémie Cohen, Thomas Counioux, Maxime Defurne, Frédéric Delmas, Philippe Delpal, Ludwik Dobrzynski, Vincent Ducas, Louise Dumoulin, Mathilde Duquesne, Alain Falou, Franck Feurtey, Pascal Garin, Olivier Gérard, Clément Germanicus, Agnès Granier, Bernard Guillien, Marc Héraud, Erick Herbin, Eva Héripré, Hélène Huard, Murièle Jacquier, Jean-Bernard Kazmierczak, Bertrand Kleinmann, Pauline Lafitte, Rémi Lantier, Sandrine Laplace, Christophe Laux, Erik Lefebvre, Gaspard and Anna Lemer, Jean-Louis Maltret, Django Manglunki, Alexis Martin, Jérôme Martin, Nicolas Millot, Elsa Montagnon, Jérémy Neveu, Van Giai Nguyen, Pierre Palasi, Bernard Pellequer, Yves Piret, Michel Reggiany, Mickaël and Ossnath Riahi, Sébastien Rychter, France Saint-Léger, Benoît Sauvage, Thomas Schmitt, Jean-Jacques Veillet, Pascal Ventura and Matthieu Villiers.

I would also like to express my gratitude to Baptiste Barreau, Nathalie Besson, Véronique Bonnefond, Dave Charlton, Yann Coadou, Thierry Depambour, Sandra Granaux, Richard Hillman, Andreas Hoecker, Eymard Houdeville, Django Manglunki, Laurent Nadolski, Bernard Pellequer, Emilien Ravigné, Bernardo Resende, Guillaume Rizza and David Rousseau for their careful reading and valuable comments. This book owes them a lot.

I would also like to thank Lison Bernet for the superb illustrations, which enhance this account of a visit to CERN.

I am very grateful to the Ecole CentraleSupélec, the Coordination des Jeunes Promos (CJP) of the Association des Centraliens, as well as the Société Française de Physique (SFP) and its Accelerators Division for their financial support. Created in 1873, the SFP is a recognized association of public utility whose mission is to represent the community of French physicists. The Accelerators Division of the SFP aims to promote research in the field of particle accelerators and to bring together academic, scientific and industrial professionals working in this field.

Finally, I would like to thank those who made the English version of this book possible: Tullio Basaglia, librarian at CERN; Elena Nash, editor at World Scientific; Richard Hillman, who helped me with the translation; and Kah Fee Ng, who oversaw the production process.

CENTRALE
Association des Centraliens

CentraleSupélec

Société Française
de Physique

Photo credit: All photographs were taken by the author